Interneteconomics.net

Springer
Berlin
Heidelberg
New York
Barcelona
Hong Kong
London
Milan
Paris
Tokyo

Paul J. J. Welfens

Interneteconomics.net

Macroeconomics, Deregulation, and Innovation

With 34 Figures and 30 Tables

Springer

Professor Dr. Paul J. J. Welfens
University of Potsdam
European Institute
for International Economic Relations (EIIW)
August-Bebel-Straße 89
D-14482 Potsdam
Germany
http://www.euroeiiw.de
and
http://www.intereconomics.net

ISBN 978-3-642-07765-4 e-ISBN 978-3-540-24762-3

Library of Congress Cataloging-in-Publication Data applied for
Die Deutsche Bibliothek – CIP-Einheitsaufnahme
Welfens, Paul J. J.: Interneteconomics.net: macroeconomics, deregulation, and innovation; with 30 tables / Paul J. J. Welfens. – Berlin; Heidelberg; New York; Barcelona; Hong Kong; London; Milan; Paris; Tokyo: Springer, 2002

Springer-Verlag Berlin Heidelberg New York
a member of BertelsmannSpringer Science+Business Media GmbH

http://www.springer.de

Printed in Germany

Hardcover-Design: Erich Kirchner, Heidelberg

Preface

The New Economy is rapidly unfolding with a dynamic mixture of computer, internet and telecommunications: the information and communication technology sector (ICT). This book is on one side a contribution to the theory of telecommunications and internet economics and on the other a regulatory policy analysis with a Schumpeterian perspective. Some crucial theoretical aspects also concern network aspects and growth theory for an open digital economy.

It may still be too early to fully understand the major features of the new economy, but the time is right to look into theoretical issues and major developments in leading OECD economies. Interneteconomics is likely to become a new standard line of research in the future. From a theoretical perspective one can already identify several interesting aspects, and some core issues relevant from an economic perspective will be picked up in this book.

Chapter A takes a closer look at the New Economy. It introduces the broader characteristics of the digital economy and addresses major problems and challenges by pointing out recent developments in telecommunications and internet dynamics in the US and Europe. Some controversial views are taken up here including the argument that there is weak evidence for productivity-enhancing effects of information and communication technologies. There are also interesting theoretical links between telecommunications policy and internet expansion. Since the telecommunications network is the basis for access to the internet regulatory approaches for the telecommunications sector will be crucial for the internet.

Chapter B present some basic facts on ICT expenditure and ICT production in OECD countries as well as new theoretical contributions to the theory of economic growth in digital open economies: Some new ideas presented are framed within a modified neoclassical model, other aspects are related to the New Growth Theory. We try to bridge here short term macroeconomic analysis and the theory of economic growth plus aspects of integration and network effects.

Chapter C looks at developments in telecommunications policy in Europe. This includes the analysis of regulatory policy issues. There are crucial problems with competition in the local loop and in other fields - even after the 1998 EU liberalization of network operation and telecommunication services. Product bundling is a new challenge which distorts not only the telecommunication market but the internet sector as well. There are several clear conclusions for policymakers.

Chapter D is one some European and transatlantic economic issues related to ICT dynamics, innovation and innovation policy. Besides theoretical aspects we raise the question whether the Euro will be negatively affected by a sustained US lead in the New Economy. We also present some crucial policy conclusions.

This partly refers to results from a research project which I conducted as Jean Monnet professor for European Economic Integration at Potsdam University in 1998/99. Furthermore my stay at the AICGS/The Johns Hopkins University, Washington DC – where I enjoyed a stimulating research atmosphere as the DaimlerChrysler research fellow in July/August 2000 – gave me the opportunity to study some important aspects of the US market. I am grateful for discussions with Andre Jungmittag in Potsdam in 2000/2001.

In the internet under www.interneteconomics.net the reader will find additional information. This innovative feature should help to generate a broader discussion about the main arguments presented in the book.

I am very grateful to Tim Yarling, Tokyo, Colleen Nagy, Washington DC, Eric Reynolds, Berkeley, Dennis Drechsler, Grenoble and Ina Brüggeman, Jaroslaw Ponder, Ralf Wiegert, Albrecht Kauffmann and Nicholas Herrmann, Potsdam, for technical and editorial assistance. It is hoped that the book gives a broad stimulus to future research in telecommunication and internet economics. The website *interneteconomics.net* has been designed by Andreas Geißler, Potsdam and myself. We hope that this website will become an interface between researchers and science and those in the wider public who are interested in ICT dynamics.

November 2001 Prof. Dr. Paul J.J. Welfens (Jean Monnet Chair)

Contents

A. Telecommunications and the Internet Boom: Analysis and Economic Perspectives

1. Trends in ICT Markets

In the 1990s the increasing use of information and communication technology (ICT) stimulated economic growth and innovation in the US and in Europe. ICT consists of information technology – computer and software – and telecommunications network operation and services. Both pillars of the ICT sector are interdependent: Cheaper and more powerful computer equipment can be used to build faster communication networks and to launch innovative digital services. Modern digital telecommunications at the same time accelerate the diffusion of knowledge and information so that there is an increasing demand for storing and processing this new knowledge/information via computers. While the US, after the splitting up of the old AT&T in 1984, has already experienced a certain level of competition in telecommunications, continental Europe and Japan have only begun to embrace competition in telecommunications in the late 1990s: 1998 was the magic date on which national operators in EU countries were forced to accept competition in network operation and in telecommunications services. The UK had already introduced competition in 1984, at which time a duopoly was introduced. This gave way to more open competition in the fixed network business after 1990. Since the early 1990s mobile telecommunications has grown rapidly in Europe, the US and Japan – the latter being quite successful with the so-called i-mode which offers mobile internet services. The internet indeed became a dynamic network of a new kind after 1990, since rather than using a dedicated line for communication it establishes communication through sending out parcels of information via different routes which are then recombined for the recipient of the message. As regards mobile telephony Europe is ahead of the US and also hopes to be a leader in mobile internet applications; third generation mobile telephony – based on UMTS technology – will offer new opportunities for fast broadband mobile services, including powerful internet services.

The dynamics of the computer market are enormous, few of the early PC innovators have survived the high growth rates in a market which was not only characterized by a high rate of technological advancement of key components – the chip and storage media – but also by the development of more powerful software. The internet plays a role in this context as computers can be ordered via the internet. They are then custom-built according to individual requests. The DELL company became a pioneer in this field with other computer companies imitating this approach. IBM, HP, SIEMENS, FUJITSU and TOSHIBA are older companies still active in the computer markets. However the many newcomers have been able to conquer large shares of the market. Computers are used in every sector, from industry to the services sector. How is the combination of faster and better computers affecting the economic system and economic development? Will the competition between the US, Japan and the EU be seriously influenced by ICT dynamics? Can the ICT sector which accounted for less than 10% of GDP in 2000 in each of these three countries (EU is considered here as a quasi-country) really change the whole economic system in a broader way? How will the macroeconomy be affected by ICT dynamics and by the internet in particular? How important is regulation of telecommunications and the internet – while telecommunication has traditionally been heavily regulated, the internet business has in contrast been relatively unregulated for many years in OECD countries.

ICT markets are cyclical markets which benefit both from cyclical expansion and from long term growth. Roughly 2/3 of US net investment in the 1990s was ICT investment, meaning its role in terms of investment is more important than the low share of ICT in GDP would suggest. High growth rates of ICT stimulated economic growth in the US in the 1990s, ICT growth was also high in EU countries, although it trailed in the larger countries – such as Germany, France and Italy - behind the US. Outside the OECD, Asia recorded very high growth in both IT expenditures and ICT expenditures in the late 1990s.

Fig. A1: Labour Productivity in High Tech Manufacturing

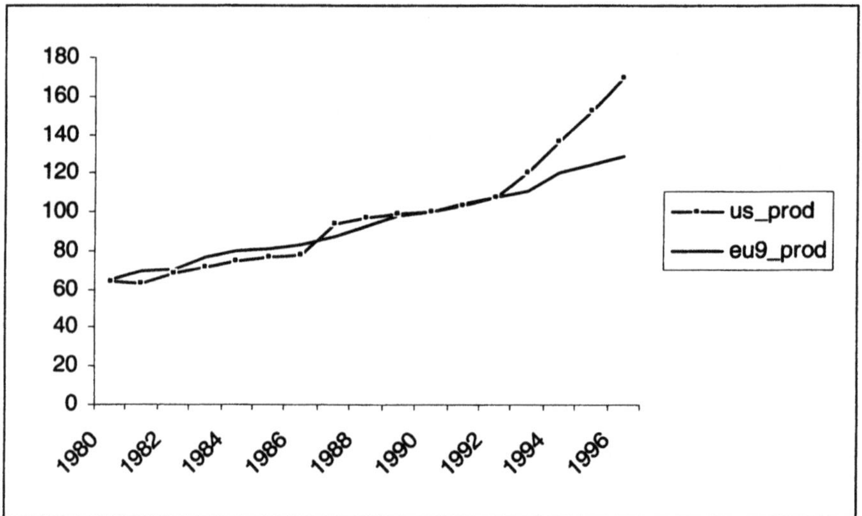

Source: RÖGER (2002)

ICT growth could lead to a New Economy in the sense that it will allow the creation of "digital markets" in which transactions are fully computerized and organized via digital networks; moreover ICT could create a New Economy in the sense that total factor productivity will be raised in a sustainable manner, the residual in growth accounting – what is not explained in terms of capital input and labor input – is thus expected to increase. A question of particular interest will also be whether ICT dynamics is mainly labor-saving or capital-saving. It is remarkable that since 1993 the US has been able to establish, vis-à-vis the EU, a strong lead in labor productivity growth in the high technology sector (Fig. A1, RÖGER, 2002).

Even more interestingly, the US achieved faster labor productivity growth than Germany in the second half of the 1990s which is in marked contrast to decades of a German lead in labor productivity growth.

Investment in IT and Telecommunications in Europe, the US and Japan

IT annual expenditure growth was close to 10% in both Europe and the US in the period 1998-2001. The structural breakdown shows that the US is leading with a share of about 45% while Europe is gradually catching up: from a share of 25% in 1998 to more than 26% in 2002. Japan´s growth rates in IT expenditure were only about half the figure of the US and Europe.

Tab. A1: Global IT Market Growth – Market Value. Mill. Euro at Constant 1999 Exchange Rates*

Million Euro	**1998**	**1999**	**2000**	**2001**	**2002**
Europe (incl. Eastern Europe*)	207.808	231.520	258.203	286.762	317.860
US	383.815	417.936	457.132	501.017	546.803
Japan	111.359	114.664	122.021	128.244	135.740
Rest of World	132.241	145.612	166.939	190.123	214.499
Total	835.223	909.733	1.004.296	1.106.146	1.214.902
			% growth		
Europe (incl. Eastern Europe*)	9,3	11,4	11,5	11,1	10,8
US	9,6	8,9	9,4	9,6	9,1
Japan	-4,0	3,0	6,4	5,2	5,8
Rest of World	15,2	10,1	14,6	13,9	12,8
Total	8,4	8,9	10,4	10,1	9,8
			% breakdown		
Europe (incl. Eastern Europe*)	24,9	25,4	25,7	25,9	26,2
US	46,0	45,9	45,5	45,3	45,0
Japan	13,3	12,6	12,1	11,6	11,2
Rest of World	15,8	16,0	16,6	17,2	17,7
Total	100,0	100,0	100,0	100,0	100,0

*includes Czech Republic, Hungary, Poland, Russia, Slovakia, Slovenia

Source: EITO 2001/POLS (2002)

Except for 1997, 2001 and 2002, market growth in telecommunications in western Europe was higher than in the IT sector (Tab. A2). With increasing mobile telecommunication penetration rates it was to be expected that market growth would slow down. Market growth is also impaired by high auction prices paid in several EU countries for UMTS licenses: Network operators have put pressure on equipment firms to cuts prices of network equipment. Growth rates of roughly 8% in IT and 6-10% in telecommunications markets in Western Europe at the beginning of the 21st century, clearly stand for above average sectoral growth.

Tab. A2: Western Europe IT and Telecommunications (TLC) Market Growth and Forecast in %, 1992 - 2002

	1992	1993	1994	1995	1996	1997	1998	1999	2000	2001	2002
IT	2,5	2,0	5,6	7,9	6,7	8,6	10,1	10,8	11,5	7,5	8,6
TLC	7,1	5,1	8,7	9,3	8,5	8,0	12,6	13,4	14,3	6,3	5,7

Source: EITO 2001 update

As regards IT expenditures per capita in (selected) OECD countries, the US was clearly leading with Euro 1635 in 2000, a much higher figure than that of the EU leader Sweden with Euro 1190. France and Germany had 2/3 of the Swedish figure – with France slightly ahead of Germany (Tab. A3). The Netherlands was slightly ahead of France. Among the large economies the UK was the leader in Europe with 906 Euro. The US expenditure on IT stood at 5.5% of GDP in 2000. In Europe, Sweden was the leading country with 4.8%, followed by the UK with 3.9%, the Netherlands with 3.3% and France with 3.2%. Germany´s IT expenditure relative to GDP was only 2.8%. Japan`s IT expenditure as a percentage has increased parallel to the dynamics in Western Europe, however, Japan`s IT-GDP ratio was of a higher level, roughly ½ percentage point above Western Europe. The higher Japanese price level for IT goods in comparison to Europe does however, lead to a slight degree of overstatement in this figure. At the bottom line western Europe´s shortfall vis-à-vis the US has been only slightly reduced in the period 1997-2000: It was 2 percentage points in 1997, down to 1.6 points in 2000. If these developments should continue, Western Europe may catch up with the US in terms of relative IT expenditure around 2010. This would not however mean, catching up in terms of IT per capita stock or in terms of IT stock relative to GDP. Stock figures reflect past accumulation dynamics, and western Europe would have to overtake the US for many years if the European IT stock-GDP figure were to catch up with the US.

Part of the US lead in IT stock figures and in IT-GDP ratios certainly reflects the relatively fast fall of relative IT prices in the USA. The more competitive US market brought sharper price reductions than the fragmented markets in the EU (with the start of the Euro in 1999 the fragmentation should, however, decline over time in the euro zone). Another reason is the US lead in the software business; the world´s leading software firms – except for SAP – are located in the US and develop new software releases in close cooperation with US firms willing to use such programs. Thus it is not surprising that according to OECD figures the

US software expenditure-GDP ratio was roughly 1 percentage point ahead of the figure for France in the late 1990s.

Tab. A3: IT Expenditures (Per Capita Figures and % of GDP)

	Per capita IT expenditure (in Euro)				IT expenditure as % of GDP			
	1997	1998	1999	2000	1997	1998	1999	2000
US	1259	1379	1498	1635	5,43	5,22	5,40	5,50
Sweden	840	966	1072	1190	3,71	4,23	4,51	4,78
UK	635	717	794	906	3,40	3,42	3,65	3,92
Japan	924	879	903	959	3,15	3,30	3,37	3,58
Netherlands	594	653	725	796	3,00	3,07	3,25	3,34
France	534	590	653	732	2,55	2,71	2,92	3,15
Finland	522	577	645	723	2,65	2,69	2,88	3,03
Western Europe	470	519	573	640	2,39	2,56	2,74	2,91
Germany	505	562	617	678	2,19	2,41	2,59	2,76
Norway	749	834	934	1018	2,55	2,85	3,00	2,74
Spain	185	218	250	284	1,55	1,73	1,88	1,99
Italy	271	289	319	358	1,47	1,59	1,72	1,84
Ireland	329	371	411	435	2,02	1,90	1,88	1,81

Source: EITO 2001

Tab. A4: Telecommunications Expenditures (Per Capita Figures and % of GDP)

	Per capita telecommunications expenditure (in Euro)				Telecommunications as % of GDP			
	1997	1998	1999	2000	1997	1998	1999	2000
Spain	411	477	576	689	3,45	3,80	4,33	4,83
Italy	454	525	617	707	2,45	2,89	3,32	3,64
Netherlands	564	646	741	861	2,85	3,03	3,32	3,61
Ireland	595	692	781	854	3,65	3,55	3,57	3,54
Sweden	629	712	791	870	2,78	3,12	3,32	3,49
UK	555	623	685	803	2,97	2,97	3,15	3,48
Western Europe	512	580	656	750	2,61	2,86	3,13	3,41
US	801	856	912	967	3,18	3,24	3,29	3,25
France	478	551	620	714	2,28	2,53	2,77	3,07
Germany	536	593	651	722	2,32	2,55	2,73	2,95
Japan	719	707	733	782	2,50	2,65	2,73	2,92

Source: EITO 2001

It also is noteworthy that among the top 20 joint stock companies there are several New Economy Firms; in the EU the New Economy is much weaker among leading stock markets, in Japan not a single New Economy firm is among the top companies. This shows that innovativeness and market growth in ICT in the US is really starting to shape crucial parts of economic output and financial markets.

As regards telecommunications, western Europe´s expenditure-GDP ratios were slightly above the US figure, probably reflecting both the European lead in mobile telecommunications and the relatively high price of telecommunication services in some EU markets (eg Spain which had the highest ratio, namely 4.8% in 2000, Tab. A4).

Tab. A5: Internet Hosts in OECD Countries

	Total hosts in OECD (incl. gTLD*)			Hosts per 1,000 inhabitants	
	Sept. 1999	Sept. 2000	% growth	Sept. 1999	Sept. 2000
United States	41,263,000	58,966,000	43	149	212
Finland	641,000	801,000	25	125	155
Iceland	28,200	35,500	26	103	126
Canada	2,389,400	3,646,800	53	77	117
Norway	396,500	512,900	29	91	115
Sweden	616,500	906,900	47	69	102
New Zealand	248,300	343,400	38	65	89
Netherlands	841,300	1,255,500	49	53	80
Australia	1,058,000	1,385,500	31	57	73
Denmark	319,500	380,000	19	61	72
Switzerland	320,200	440,400	38	44	60
Austria	241,200	437,400	81	30	53
United Kingdom	2,076,500	2,956,700	42	35	50
Belgium	290,100	390,500	35	28	38
Luxembourg	8,100	13,600	68	19	32
Japan	2,489,500	3,931,800	58	20	31
Germany	1,713,300	2,443,100	43	21	30
Ireland	62,400	111,300	78	17	30
Italy	584,200	1,695,000	190	10	30
France	773,700	1,105,700	43	13	19
Hungary	116,200	151,100	30	12	15
Spain	385,300	604,400	57	10	15
Czech Republic	107,100	129,500	21	10	13
Greece	75,200	130,900	74	7	12
Portugal	69,700	121,500	74	7	12
Korea	320,100	481,500	50	7	10
Poland	163,400	295,500	81	4	8
Mexico	198,600	348,400	75	2	4
Turkey	85,400	206,000	141	1	3

Source: EITO (2001)

In 2000 the host density in the US reached 212, Finland was No. 2 with 155 followed by Iceland, Canada, Norway and Sweden - all exceeding 100 hosts per 1000 inhabitants. Among large EU countries UK was leading with 50, followed by

Germany, Italy, France and Spain. The supply side of information in the net thus clearly is dominated by the US and North America, respectively.

In 2001, the number of internet users per 100 inhabitants was 55 in the US, but only 30 in Western Europe and 32 in Japan. Among the large European economies the UK was leading with 42, followed by Germany with 37 and Italy and France with 29 and 28, respectively. In 2003 the US will – according to POLS (2002)- still have a clear lead with a user density of 2003, followed by the UK with 57 and Germany with 49, narrowly followed by Italy and France (Tab. A6). The Scandinavian countries – not shown in the table – will maintain their clear lead in terms of user density at the beginning of the 21st century.

Fig. A2: Worldwide Number of Internet Hosts, Millions, 1995-2000

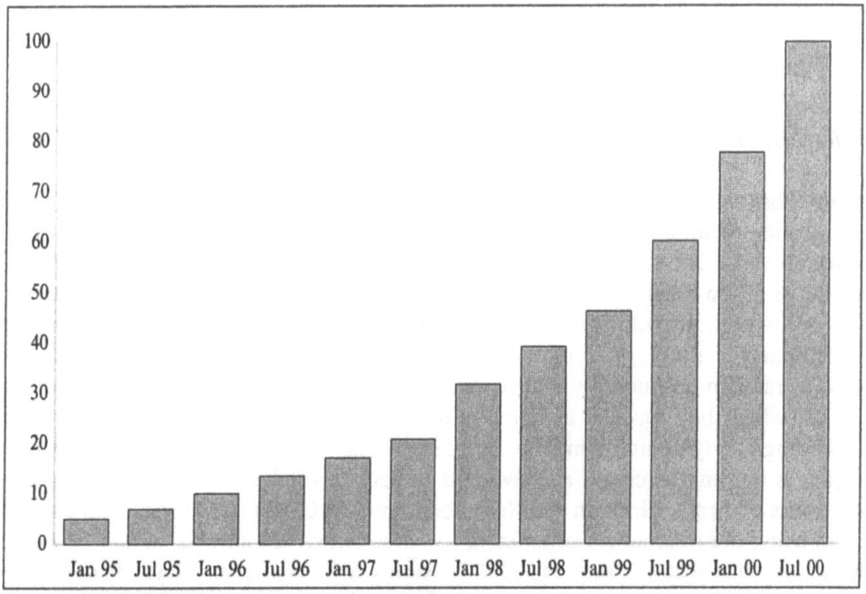

Source: EITO (2001)

Tab. A6: Internet User Density (users per 100 inhabitants)

	1998	1999	2000	2001	2002	2003
France	7	11	17	28	35	42
Germany	12	15	28	37	43	49
Italy	4	6	19	29	38	46
Spain	5	9	14	24	32	39
United Kingdom	15	22	29	42	50	57
Western Europe	8	12	21	30	37	43
USA	29	38	48	55	61	66
Japan	11	14	21	32	40	47

Source: POLS (2002)

Tab. A7: Mobile Subscriptions per 100 Inhabitants

	1998	1999	2000	2001	2002	2003
France	19	35	51	66	80	93
Germany	17	28	58	80	92	102
Italy	35	52	74	87	95	100
Spain	18	38	62	84	94	100
United Kingdom	22	41	68	85	97	103
Western Europe	21	36	58	74	86	94
USA	25	31	40	51	62	73
Japan	31	38	47	58	70	81

Source: POLS (2002)

2. Towards a New Economy

A Revolution in Telecommunications

Many observers of the incredible technological progress in information and communication technology (ICT) consider the leading OECD countries as being characterized by a New Economy. A core element of the New Economy is the ICT sector itself: on one hand the telecommunication sector, which is above all characterized by high growth rates of patents since the 1990s, and on the other hand the information technology sector. Computer production has recorded enormous growth rates in productivity and software expenditures exploded in the US in the 1990s, when they reached a figure as high as spending on R&D. On top of the technological dynamics came the spreading of the commercial use of the internet, which is helping to create a networked society beyond the traditional computer networks of firms. Firms in the New Economy (eg CISCO or DELL) can benefit from using the internet for organizing its value-added chain. One cannot rule out, however, that sectors from the Old Economy will also reap major benefits by reorganizing the value-added chain in an innovative way.

There is little doubt that US growth and labor productivity have increased in the 1990s. The growth rate of labor productivity in 1995-2000 reached 3.1%, while in 1973-95 it was 1.4% p.a., and the growth rate of potential output accelerated in the 1990s while that in Japan reduced. In Euroland there was only a modest increase (COUNCIL OF ECONOMIC ADVISERS, 2001). The US acceleration is, however, not as impressive as it looks at first glance, since the US is using hedonic deflation methods for computer equipment which strongly increases real value-added of the computer-producing sector, above all in a period in which the computer sector increased. In the EU only Sweden, Denmark, and France used hedonic deflation methods in the 1990s. Moreover, since 1995 firms in the US include expenditures on software as investment, which is a major change in the system of national accounting (put simply: output is the sum of consumption plus investment). The US does not show a sustained increase in labor productivity in the second half of the 1990s based on traditional concepts of accounting, rather there is an increase in 1995/96 and a slight decrease in labor productivity growth in 1997/98, which is the normal pattern at the end of the business cycle. The statistical increase in labor productivity is magnified if one additionally takes into account the effect of hedonic deflation concepts (UPPENBERG/VANHOUDT, 2000, p.13); reclassifying software expenditures has increased US GDP growth in the second half of the 1990s by slightly more than one percentage point, and if the EU would follow a similar procedure the effect for the Community would be roughly the same. Note that the idea of using a hedonic price index is that the quality aspects of a product contribute to customer's willingness to pay but that process innovations reduce the price over time, and while using conventional deflation methods for computers would indicate a decline of computer prices by about 20% in the period 1993-99 in the US, the use of hedonic deflation concepts, taking into account that new computers come with fast processors, larger hard disks etc., would indicate a fall by 80% (UPPENBERG/VANHOUDT, 2000). Since computer production is a growing sector in the US the switch to hedonic pricing indeed increasingly affects the deflation of nominal GDP figures over time. To put it differently, the real growth rate in the US has partly increased in the 1990s as a consequence of new statistical procedures. Nevertheless, there is little doubt that the transatlantic growth differential is considerable even if we correct for the asymmetric statistical changes.

The basic observation of the 1980s and the 1990s is that information and communication technology have increasingly contributed to high growth in the US, while the EU (or Euroland) has been lagging behind in ICT dynamics. The EU countries lead in the 1990s only in the field of mobile telecommunications, as they benefited from choosing a uniform GSM platform which helped to exploit network effects in a rapidly growing new market. Taking a look at the use of the internet, however, there is a clear lead of the US compared to the core of Euroland, that is Germany, France, and Italy. The US lead in computer density where the figure was 2:1 for the US relative to Euroland in 1999 together with a rapid commercial exploitation of the internet in North America seems to contribute to a sustained transatlantic growth differential in favor of the US and Canada. In the following analysis we will not only look at ICT dynamics on both sides of the Atlantic, we also will highlight some basic differences in the regulatory structure of the telecommunications sector which accounts for about 1/3 of the overall ICT sector.

The commercial internet was started in 1993 when the National Science Foundation (NSF) decided to leave the management of the backbone fully to competing, commercial backbones which quickly raised the question of interconnection of commercial backbone networks. In 1986 the NSF funded the NSFNET (56 kilobit network,) which was a network established to enable long-distance access to five supercomputer centers in the USA. As a commercial network of networks it became increasingly popular in the private sector in the 1990s: e-mail, www services, and e-commerce became increasingly popular in the private sector.

The internet is a network of networks with a light set of rules: national and international regulation on voice telephony is not valid for data traffic over the internet. Within less than a decade a transmission network and new protocol for data transmission has conquered the world; the formerly government-owned US internet system has become a worldwide net which is open to all countries, every firm, and every man/woman on the globe. What is so special about the internet and the so-called new economy? Which countries and firms will be the leaders in the global internet society? What policy challenges emerge for Europe, the US, Japan and the NICs, and the developing countries?

A specific feature of the internet society is the networked economy in which access to and transmission of information is relatively easy. There are some specific economic aspects of a network, many of which are known from the introduction of railway and highway systems. The more users are linked to the network the greater the marginal benefit for the representative user, as long as there is no congestion. The telecommunications network, and the modern internet system. has, however, additional aspects. It allows mutually beneficial exchange of information and the flexible recombination of knowledge and know-how. This is all the more important in the knowledge society, the modern society which emerged in the 1990s when knowledge as a distinct production factor became increasingly important. Modern digital networks, in particular the internet (based on the Internet Protocol), facilitate the cheap, effective, and flexible combination of knowledge. The fact that the wage ratio of skilled to unskilled workers has increased in the US and many other OECD countries indicates the increasing economic role of knowledge. Moreover, the US has experienced a strong growth in patents granted in the 1990s, in particular in the second half of the 1990s; a court decision indicating the patentability of computer software might have stimulated the patent surge (COUNCIL OF ECONOMIC ADVISERS, 2000a, p. 122).

While real capital became an increasingly important input factor in the 19th century and the first half of the 20th century with the spreading of highly capital intensive economies of scale products, the second half of the 20th century was characterized by strong relative growth of software and knowledge as inputs in an increasingly flexible production process of goods and services. The long term growth of knowledge as a factor input is reflected in rising shares of well-educated workers and employees, as well as rising R&D/GDP ratios in OECD countries and NICs. Flexible knowledge-based production has been enormously facilitated by the growing application of computer and PC technology in combination with ever faster transmission of information. The US Federal Communication Commission had already issued regulation referring to computer-based data transmission (enhanced services) in the 1980s. In the US data transmission was subject to lighter rules than voice telephony in the sense that no access fees were charged for data transmission. Nobody could anticipate that this rule would be extremely beneficial for the growth of the internet, which is basically digital data transmission based on a special internet protocol. There is a rapid growth of ICT markets in Europe, the US, and some other regions. In 2000 overall global market volume is close to 2000 bill. DM, where half of the volume is for telecommunications. In the global ICT market the US, Europe, and Japan represented 35, 31, and 11% in 2000, while the US had a clearer leadership in information technology markets. The US accounted for 44%, Europe for 29%, and Japan for 11%; by contrast Europe´s share exceeded that of the US in telecommunications. Europe represented 34%, the US 26%, and Japan 9% in 2000; growth rates in the telecommunications market in Europe will be 11%, 8% in the US, and about 6% in Japan. High growth rates are expected in network infrastructure (+25.6%), equipment (25%), and services (9.4%) in Europe (EITO, 2000b). Growth rates in Western Europe´s net infrastructure are expected to increase to over 10 % in 2001, up from 6.8% in 1999. DSL network expansion is crucial for incumbant operators in all leading EU countries. Deutsche Telekom AG, which has already reinforced its market position with 7 million internet (T-Online) users, had 300 000 DSL customers by the end of October 2000, and expects two million by the end of 2001.

The interaction of telecommunication liberalization and the expansion of internet is likely to stimulate economic growth (WELFENS/JUNGMITTAG, 2001). As regards Germany almost 95% of all households could get access to asymmetric DSL by the end of 2002, which is a kind of turbo for the existing fixed link network, making transmission much faster than in the old network. Similar figures will hold for the US and most EU countries around 2003, so that there will be an acceleration in data transmission speed. DSL will partly compete against leased lines as soon as symmetric DSL offering upstream and downstream up to 2 Mbps is available. This will put pressure on operators which offer leased lines to come up with new services and lower prices. The corporate telecommunications market will strongly benefit.

There are four major developments which charatecterize modern telecommunications – and links between telecoms dynamics and growth - in the 1980s and 1990s:

1. Rise of innovation dynamics: In the 1990s there was an increasing dynamic in telecommunication technology which is visible in the rapid growth of telecommunication patents. It trippled from about 2500 granted annually in 1990-92 to 7674 in the US in 1998 (FCC, 2000). The European Patent Agency shows a similar development, where part of the new technological dynamics can be attributed to (anticipated) privatization of the national fixed network operator in EU countries: As the management was facing initial private offering of stocks, there were strong incentives to improve long term revenue and profitability, and here accelerated innovation obviously is important. Innovations include faster transmission of data, new switches based on the internet protocol, improvements in computer technology and advanced software, and fibre optical cables (which will allow for enormous increases of data flows at low costs), all of which play an important role in the innovation race in the communication industry. The rise of innovation dynamics should reinforce telecommunication´s traditional lead with respect to labor productivity: It rose an average 6% per year in the period 1951-97 in the US, while average labor productivity in other sectors hovered around 3.5%; the relative price of telephone services has fallen strongly since the late 1980s, when telephone services prices even fell in absolute terms (FCC, 2000, p. 5-1, p. 13-4). Households have spent about 2% of their expenditures on telephone in the period 1980-1998, where the share increased from 1.9% in 1980 (and 1985) to 2.3% in 1998 (FCC, 2000, p. 4-3).

2. Digital transmission and digital equipment has created better quality and faster transmission on the one hand, and it has erased the demarcations between markets on the other: Technological convergence, that is digitization of signals (voice, data and images), has not only created a larger digital multimedia-TV-entertainment market, it has reinforced competition (at least transitorily), as well.
3. Rapidly rising telecommunications (the information exchange mainly in the business community) at the national and international level has a potential impact on diffusion of knowledge and best practice technologies. JUNGMITTAG/WELFENS (1998) found empirical evidence for Germany that use of telecommunications was significant for growth accounting in Germany. Some examples are a higher share of international phone calls, which partly reflect the rise of trade, international investment and tourism, not to mention genuinely international trade in information. After the early 1980s there was a rapidly growing internationalization of telephony in the sense that the number of international phone calls has increased strongly. A comparison of intra-US calls to intra-EU calls still shows much higher prices in Europe than in the US. In 1980 there were 200 million calls made from the US, in 1998 there were 4.5 billion. Carriers billed 60 US cents per minute for international calls in 1998, which is a decline of more than 50% since 1980. The lion´s share of US billed minutes by country is represented by Canada, Mexico, Germany, UK and Japan which accounted for 15.8, 12.3, 5.7, 4.8 and 3.3%, respectively. In western Europe strongly falling international prices were recorded only after the 1998 deregulation. The price asymmetry between the US and the EU (for in- versus outbound transatlantic calls) could partly explain observed growth differentials, to the extent that international flows of information contribute to economic growth. The traditional system of accounting and settlement rates administered by the International Telecommunications Union (ITU, Geneva) contributed to maintaining rather high international telecommunication tariffs into the late 1990s, when new private operators entered various EU markets in the context of deregulation and privatization.
4. There is growing competition in international telephony since the mid-1980s: Since 1985, when MCI first entered the market for international telephony in the US, dozens of carriers have started to offer international telecommunications services based one their own facilities or on leased lines from other carriers. In the EU it was only after the introduction of competition in telecomms network operation and telecomms services in 1998 that competition within EU countries increased, where newcomers from abroad who were facing shrinking markets at home in the new open market setting, as well as mobile operators and cable TV, contributed to overall competition in the telecomms sector at large; this sector´s growth is strongly linked with the use of advanced computers and modern digital network technology. The rapid growth of leased line sales in the US and Europe stands for a rather efficient way of optimally using existing networks.With fast DSL (digital subcriber line) service this can be taken one step further, as a given line can be split into voice

telephony and DSL – we note here that DSL is a dedicated line and thus should be slightly better in terms of quality than using cable TV network, which can also provide broadband transmission: The result is fast internet transmission which includes options for streaming TV content.

Fig. A3: Telecommunications Patents Granted in the US, 1980-1999

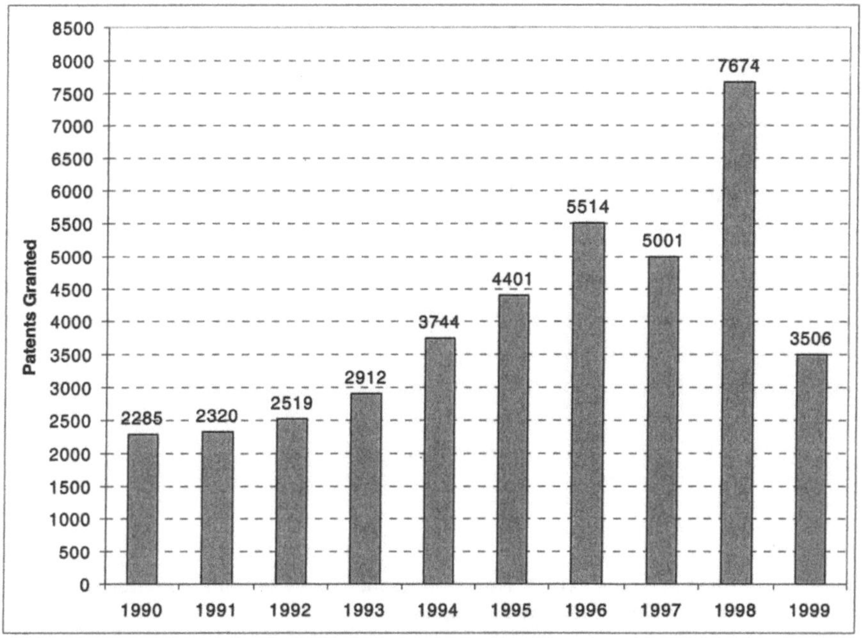

1996 total reflects one-time change in law affecting patents.
1999 data reflects patents as of June 30, 1999.
Source: U.S. Patent and Trademark Office, Technology Profile Report – Telecommunications, Classes 370, 375, 379 and 455 / Source: FCC (2000) Trends in Telephone Service, March 2000, Washington DC.

2.2 Growth, ICT Dynamics and Innovation

Growth rates of real GDP in the US have shown a remarkable increase over the period 1991-2001. The average growth rate in the US during the period 1993-2000 was almost 4%. A marked acceleration occurred in the period 1995-2000. By contrast, growth rates in Japan have declined in the twelve years since 1988. In the EU there has been a modest growth acceleration in the decade since 1993; Germany and Italy were largely responsible for weak growth since both countries had growth rates below the EU average. In addition to considering country perspectives one also needs, of course, to focus on investment dynamics. Given the fact that more than ½ of US investment growth in the 1990s was related to ICT, it is important to recognize that the EU in the seven years since 1993 had indeed lower growth rates of investment than the US, and part of this gap (and the lower growth rates of GDP) – see the following figures - is due to its much lower levels of ICT investment respective to the US (RÖGER, 2002).

Fig. A4: Growth of Real Gross Domestic Product in the European Union, the US and Japan, 1984-2002

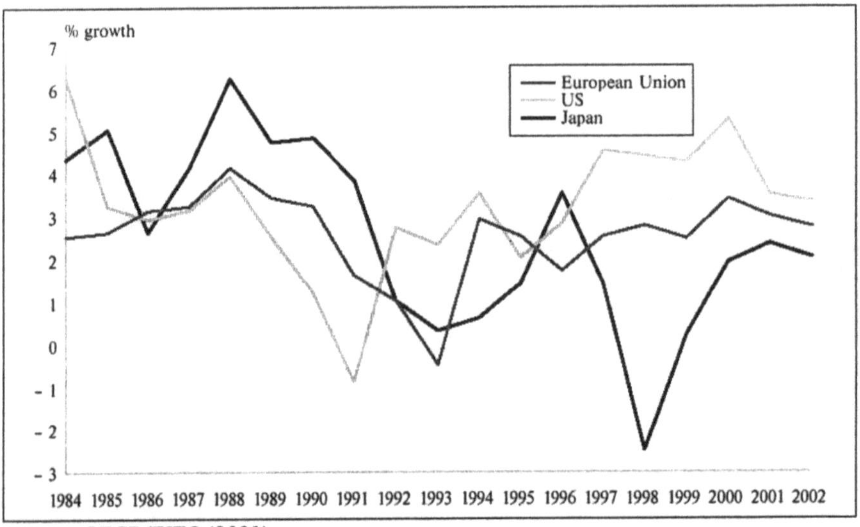

Source: OECD/EITO (2001)

Fig. A5: Growth of Gross Private Non-residential Fixed Capital Formation in the European Union, the US and Japan

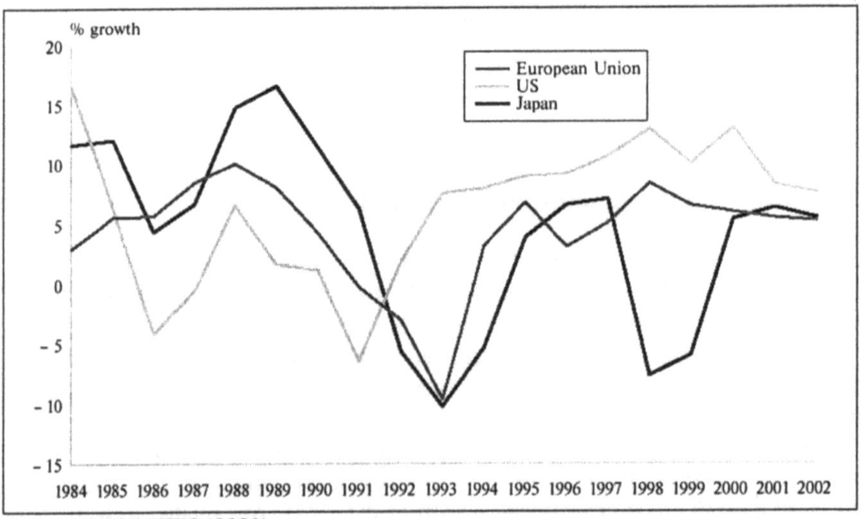

Source: OECD/EITO (2001)

As regards US economic growth, the US COUNCIL OF ECONOMIC ADVISERS (2001, p.28) has divided labor productivity growth into the periods 1973-95 and 1995-2000; it found that productivity growth increased from an annual average of 1.39% in the first period to 3.01% in the second period, where capital formation (without ICT capital) contributed 1.09 points in the period 1995-2000, up from 0.70 in the period 1973-95; ICT capital formation contributed 0.41 in 1973-95, but 1.03 points in 1995-2000. Structural total factor productivity growth increased from 0.40 in the first period to 1.59 points in 1995-2000 where the impact from the computer sector alone was 0.36 in that period (up from 0.18 points in 1973-95). What is remarkable here is that the contribution of ICT capital formation and total factor productivity growth have increased strongly. The share of ICT investment increased from 3% of GDP to 6% in the US in the 1990s. It is true that computers, even in the 1970s and 1980s, already played a crucial role in investment, however, it seems that learning costs limited the productivity effects of modern computer equipment in the two decades prior to the 1990s. It is also noteworthy that the share of the computer sector (including software) in GDP increased from 2.9% in 1990 to roughly 5% in 2000. This expansion was without doubt stimulated by the rapid fall of relative computer prices – falling by about 25% annually in the late 1990s in the US - as measured by hedonic price indices.

An important aspect of the increasing use of ICT products in many sectors of the economy – including the still dominant (as measured by output and employment shares) Old Economy – is that combining computer power and the internet allows firms to relatively easily realize three changes simultaneously:

- Raising the rate of product innovations and of product differentiation.
- Producing more customer-tailored options, allowing firms to charge a higher average price, ie. to move from a competitive market to a market with monopolistic competition.
- To reduce manufacturing costs and marginal costs, over time.

There are two other aspects related to ICT expansion:

- In addition there could be positive network effects; such network effects are supported by open standards. These played a considerable role in ICT in the 1990s.
- Moreover, there are static or dynamic scale effects.

The microeconomic perspective in relation to the first three points is shown in the following diagram in which the initial cost curve is k'_0 and the original demand curve is DD_0. The initial market equilibrium point is E_0, at which point the market-clearing price in the competitive environment is equal to marginal costs.

Now assume computers plus the internet are introduced so that product innovations can be launched faster and product differentiation becomes more easily possible. Product innovation means that the demand curve becomes steeper (DD_1). As more varieties are produced there might be somewhat higher costs – an effect we have neglected in the diagram. Instead we emphasize the impact of product variety on price setting, namely that the market in the short run switches to full monopolistic competition resulting in an increase in the average price from p_0 to p_2 while output is raised to q_2 (the rightward rotation from DD_1 to DD_2 reflects consumers´ increased willingness to pay in the presence of more varieties/

innovations realized by innovative built-to-order manufacturers); and since ICT equipment allows cost cutting in the medium run we have a fall in the price to p_3 in conjunction with another increase in output (q_3). Indeed, a highly significant aspect of the internet is that it allows firms to produce built-to-individual order in a very straightforward manner. This has been amply demonstrated by computer firms.

Fig. A6: Effects of the Use of ICT in Markets

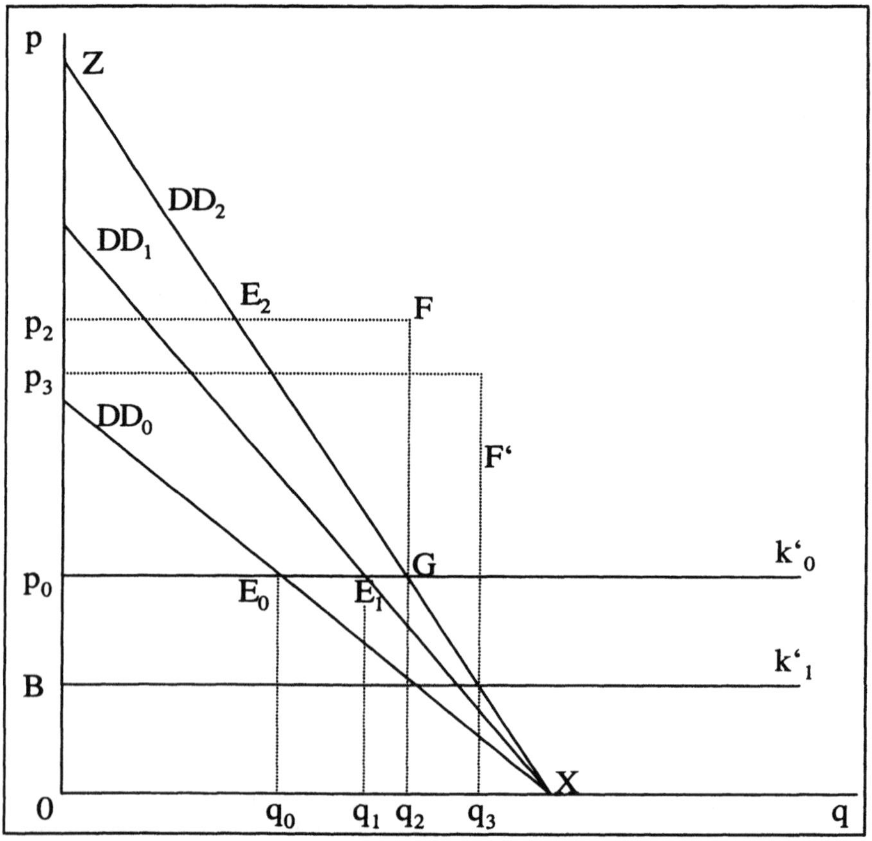

ICT dynamics are to some extent associated with the influence of standards in the digital sector of the economy. Eg Ericsson in association with some other companies set several new standards in the 1990s. This implies that there will be network effects for those capital goods producers which offer equipment tailored to the new standard. The network effects imply a rightward shift or rotation of the demand curve. If the capital equipment firms (or the software firms) considered, are facing dynamic economies of scale, a second effect associated with the demand-side network effects is that the marginal cost curve will shift downwards so that the new equilibrium is E_1 with a higher quantity and a lower price compared to the original equilibrium E_0.

Fig. A7: Network Effects and Dynamic Scale Effects

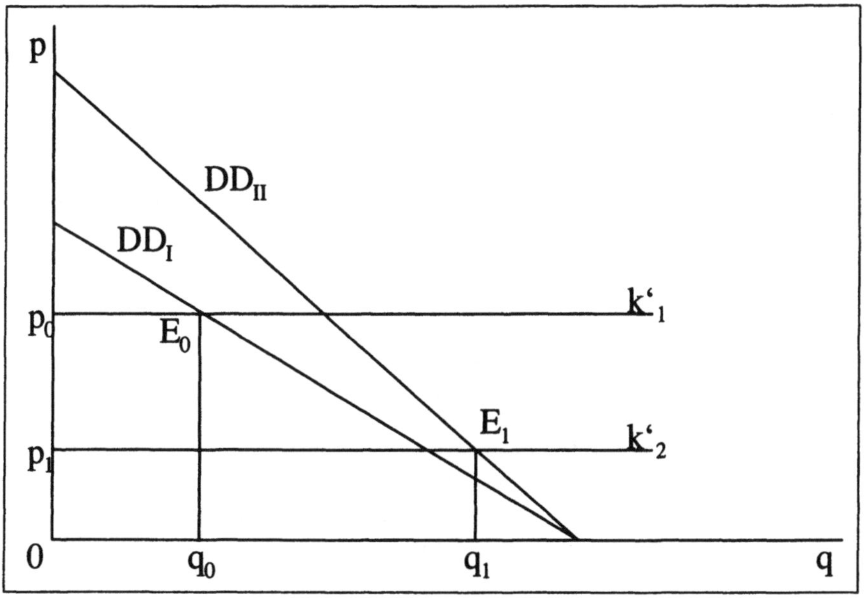

2.3 New Economy

Higher innovation dynamics, falling prices for long distance, international telecommunication, and more intensive competition all amount to a growing economic role of information and communication technology. ICT technology comprehensively applied to the economy has created a „New Economy" whose core consists of the following elements:

- A networked society based on digital technology which allows very fast transmission and effective data storage. Network effects mean that the marginal social utility of individual investment in a network is higher than private marginal utility. To put it differently, there is some self-acceleration in sectoral demand as networks expand, especially in the medium term network expansion stage, which corresponds to the medium arch of a logistical curve (for sectors with network effects the logistical expansion path is steeper than for non-network sectors). Indeed, digital technology means enormous progress over the former analogue telecommunication technology since one can build faster, more powerful, and more flexible networks. METCALFE`s Law (ie see Newton´s Telecommunications Dictionary), which says that the value of a network increases in proportion to the squared number of users of a network, is a specific expression of network effects.

- There is considerable capital-saving technological progress in the computer & internet society. Chips and computers play an increasingly important role. MOORE´s law, stated by Intel´s cofounder George Moore, says that the density of circuits we can place on a chip doubles every eighteen months with hardly any significant increase in costs. This law implies that computers in 2000 are about 66 000 times more powerful, in terms of processing power, than the typical computer of 1975. Therefore, computer-intensive products should fall in relative prices over time. However, there is a caveat, namely network effects; there will only be a weak fall in relative price (or even an increase) if the respective computer-intensive product or service shows strong network effects on the demand side.
- There is a considerable underestimation of the investment GDP ratio in OECD countries in the computer and internet society. The use of software, traditionally not counted as investment, became more important in OECD countries in the 1980s and 1990s. The US is a global leader in software investment with spending on software running slightly above 2.5% of GDP in 1998; leading EU countries reached just half this figure. It is remarkable, both with respect to software expenditures and with respect to computer expenditures, that not only industry is spending heavily on these two items. Households are also investing in computers and software (read computer literacy) which is welcome by firms using computers. The implication is that there is considerable underestimation of investment in OECD countries.
- Knowledge is playing an increasing role at the beginning of the 21st century. It reflects the accumulated effects of 120 years of public schooling and growing university education in Europe, the US, Japan, and part of the developing world on one hand, while it reflects more than a century of industrial research and development efforts conducted in a professional manner with the incentives provided by patent protection on the other.
- The diffusion of knowledge, with the exception of tacit knowledge, is accelerating in the internet age. This means that the „production" of knowledge should have larger productivity effects. Quite crucial are firm-internal intranets which create high-speed avenues for the diffusion of knowledge and the organization of cooperative knowledge exploitation.
- The digitization of information and images, in combination with novel storage and streaming technologies, has created revolutionary ways in enhancing the tradeability of services; digital services are more personalized and they are increasingly ubiquituous due to mobile telecommunications. With 3G technology (operating around 2002), and the already existing Japanese mobile i-mode, the capacity of data transmission via mobile phones is almost as high as with standard fixed telecommunication networks. This is important since mobile telephony is still expanding rapidly at the beginning of the 21st century. In June 1987 there were 0.9 million cellular telephone subscribers in the US, in June 1995 there were 28.2 mil-

lion, and the June 1999 figure stood at 76 million (FCC, 2000, p.2-3). In 2005 it will be roughly 150 million in the US, with most phones using digital technology. In the EU mobile telephone density is around 40% in 1999 and will be close to 75% in 2005, where almost all mobile phones used digital GSM technology at the turn of the century.

The growth prospects of countries leading in the internet should clearly improve in the long run so that a new long wave might have started in OECD countries. There will be countries such as the US which grow strongly because they are leaders in internet technology generation, as will others (ie Sweden, Japan) which are leaders in applying fixed and mobile internet technologies. Part of the differences in economic growth will stem from differential abilities to create .com firms, that is the characteristics of the financial market are important, too.

2.3.1 The Digital Network Economy: Ten Basic Elements of the New Economy

1. The New Economy means an economic system of networked computers and advanced telecommunications services, including novel value-added services which have become possible due to digization of communication. Much more complex computations and projects became possible with the dramatic fall of computer chip prices in the 1980s and 1990s, which is a powerful stimulus for innovation. Combining this with modern telecommunications creates a new window of opportunities for the intelligent integration of existing and new knowledge. ***In the new economy there is more innovation and faster diffusion.*** Patent applications have strongly increased in the US in the second half of the 1990s. Civilian R&D-GDP ratios in OECD countries and NICs will further increase in the long run.
2. The structure of the Old Industry is changing as networking is easier in the computer and internet society. This change will create the possibility to exploit new opportunities for outsourcing. The creation of so many new internet portals in industry indicates that there are declining transaction costs in e-markets. ***The role of markets (and SMEs) is reinforced in the New Economy.***
3. One of the biggest winners in the new economy is the chip and computer industry, which has produced astonishing technological progress strongly linked with miniaturization techniques. ***The new economy is partly capital saving*** so that a smaller investment output ratio will suffice to replicate the long term growth rate of the new economy result, but this does not necessarily mean that the return on investment is falling. The main reason for the absence of this phenomenon is that the core sectors of the new economy all benefit from either economies of scale or of network effects (in the sense that the marginal utility of existing users of a network is increased as new users are linked to the existing net). If the new economy is capital saving one may anticipate a birth rate of new firms which have to put up relatively small amounts of capital to enter a market. However, there are obvious differences

between the USA and UK on the one hand, and continental Europe with its universal banking system on the other.

4. Introducing digital telecommunications does not only mean technological progress in telephony. It also has powerful pro-competitive effects since digization means that existing market demarcations for transmission of voice, images and data are no longer valid. ***Market entry becomes more open and price competition more intense, which might reduce inflation pressure.*** The natural reaction of firms, beyond plans for mergers and acquisitions which often meet resistance of competition authorities in the US or the EU, is to try avoiding the intensified price competition in markets for standardized products. Hence firms will launch more differentiated products and product innovations in order to restore adequate profitability. This reinforces our first point (see above).
5. The internet as one element of the New Economy means that consumers (and the demand side in industry) are both more informed and more flexible in purchasing. Many demand curves will become more price elastic, and this will hold in particular for goods which are easy to categorize so that they are suitable for e-commerce - very complex products can hardly be sold online. Price mark-ups of standard products should fall and profits of firms in the respective industries will fall; the consequence should be additional mergers and acquisitions through which firms try to restore their initial market power in the enlarged e-market. Thus the new economy contributes to an ***acceleration of foreign direct investment.*** This is reinforced by the creation of many new internet-based firms which go public if they have survived the first critical years: the M&A menu list gets longer and more interesting as the new economy unfolds.
6. The new economy is thus more market-based, more competitive, more innovative, and more international. ***The new economy could also be more unstable than the old economy*** because there is a tendency of more homogeneous information in the world economy, with most all leading market actors having almost instantaneous access to similar information (could be correct or incorrect). Herd behavior might thus become more common. Moreover, the differential in the speed of adjustments in financial markets and product plus labor markets will increase. As is well known from the Dornbusch model of exchange rate, determination differences in adjustment speeds in the financial and real sector lead to overshooting phenomena so that price fluctuations in financial markets could increase.
7. Digitization would allow the replacement of public TV, a kind of collective good with zero marginal costs of transmission, with private customized TV. Therefore the digital economy creates a new entertainment industry on the one hand, and on the other hand it amounts to establishing a fragmented audience for politicians who could find it increasingly difficult to address a large audience. ***In the new economy society becomes more difficult to influence politically,*** which creates problems for countries with major internal conflicts and lack of basic political consensus. This also could be a problem if political consensus-building would be at a premium; it might indeed be needed to suc-

cessfully cope with the potential for increased instabilities in the world economy.

8. Hardware is not useful without intelligent software which increasingly stands for a high share of investment expenditures (broadly defined) in OECD countries. Combining internet and new software means that the ***new economy will have a global platform for fast innovation and fast diffusion.*** Speed and internationalization matter a lot in the internet economy. Unclear copyright and patenting problems could, however, undermine the potentially high growth of the increasingly software-based digital economy.
9. At the same time the demand for knowledge and skilled labor is increasing. There will be a relative increase in the wage of skilled labor which encourages training and education efforts. The internet is a powerful basis for supplying broader and better training and education programs, however, e-training and e-university have only just appeared at the turn of the millennium. ***In the internet society the education sector will be booming, and countries with efficient university systems should benefit from extra growth (countries underinvesting in university education and retraining will fall back).*** There is one important caveat: If the internet contributes to reducing tenure of employees in firms, this will reduce the incentive for companies to invest in training and company-internal education.
10. The ***new global economy will be characterized by a rising global economic divide***. Even if one were to donate many computers, mobile phones, and software packages to poor developing countries, their high illiteracy rates make exploiting the internet impossible. The big winners will be OECD countries and certain newly industrializing countries, as well as those developing countries with a strong education base and a liberal trade (and investment) policy. The internet network offers a potential which has to be exploited carefully, and here the global, international and national divide is beginning. The internet requires literacy which is a problem for 50% of the adult population in developing countries, and, to the extent that the internet raises factor productivity, the losers of the internet age are easy to predict. This also points to a renewed challenge for much higher education expenses in developing countries. Literacy is one important requirement for the exploitation of the internet. Developing countries can catch up only in the long term because lack of basic schooling cannot be remedied in the medium term.

The problem of a rising international economic divide might be accentuated by the empirically documented fact that spillovers from R&D activities tend to be geographically localized (JAFFE/TRAJTENBERG/HENDERSON, 1995). Moreover, AUDRETSCH/FELDMAN (1996) have shown that spatial concentration of innovations was significantly higher in sectors where knowledge generation, as proxied by industry R&D/sales, the importance of academic research and the use of skilled labor, was relatively crucial. If the internet society will be mainly a knowledge-based economy, the above argument leads us to expect that there will be regional clusters of ICT. This hypothesis neither rules out that there could be new economic centers of gravity emerging nor that there could be new patterns of innovation, ie patterns strongly exploiting the network characteristics of the internet society. Countries which fail to develop an ICT cluster region will suffer from falling terms of trade. Therefore economic convergence might become more difficult to achieve in the 21st century. A counter-argument could emphasize the diffusion-enhancing qualities of the internet, but this argument is hardly valid in a North-South context, rather it might be valid for the relation between OECD countries and Newly Industrializing Countries. Some NICS, ie Singapore, Taiwan and Malaysia, have launched massive efforts to assume a leading role in the internet society.

Problems with Information Industries in the EU

The EU has various problems in regards to the collection and dissemination of data at reasonable prices. The European Commission itself (FT, 2000) has noted that the EU could narrow the lead of the USA in information industries by removing barriers to the commercial exploitation of information generated by the public sector. The Commission has argued that access restrictions and high costs have limited the economic contribution of public sector information in the EU to 68 bill. Euro annually, just about ½ the US level, where open information policies have stimulated the growth of a more dynamic market. The Commission's Report, based on research by Pira International/University of East Anglia and Knowledgeview, has suggested that that EU governments should make public sector information available at marginal costs, preferably in digital form. The report says that EU governments invest 9.5 billion Euro annually in public sector information, compared to almost 19 bill. in the US. Serious data problems in the crucial ICT sector are certainly a formidable challenge for policymakers and the business community, since decision-makers face the risk of inadequate policy and investment decisions if needed data are not readily available. While the UN, the EU and the OECD have contributed to harmonized data collection in key fields of the system of national accounts and some related fields, ICT is a relatively new field which has been neglected in official statistics. With the support of the Commission EITO has started to regularly collect key data on the ICT sector. Certain ICT data are also collected by large private firms, ie IBM and International Data Corporation. Data sources have to be systematically identified and problems of generating comparable data sets must be studied carefully. Special data sets might also be compiled with the help of the Commission and other agencies.

Europe is lagging behind the US for several reasons: One interesting aspect concerns computer prices. Computer prices in the US are much lower than in the EU – probably by 30% in 2000. The price of telecommunications and the internet is lower in the US than in most EU countries. Regulatory restrictions on retail discounts are part of the problem in Germany. While major US telecommunication companies are developing new products and net-based services for the US (which is characterized by fierce competition) and world markets, the EU is a largely fragmented market in the field of fixed network telecommunications. Only in mobile telephony can one find a tendency to create a pan-European network, with Vodafone being the leader in this market. While there is a single market in the EU on paper, reality is different. Roaming charges in the EU are expensive as mobile users have to pay for every call from EU member country i to j, and both the calling and the receiving party have to pay.

2.3.2 Some Reflections on Computers

The US has experienced highly sustained growth in the 1990s, which is a mild contrast to the modest growth of the 1970s and early 1980s. Researchers (e.g. OLINER, 1993; OLINER/SICHEL, 1994; SICHEL, 1997; 1999; OLINER/SICHEL, 2000; STIROH, 1998, 1999) have asked whether productivity growth in the 1990s was adequately measured in an environment in which the number and power of computers increased strongly. Besides leading researchers, the Council of Economic Advisers and the Federal Reserve System have conducted research focusing on productivity increases and the role of information technology in recent US growth (see also the Economic Report of the President, 2001). Computer densities differ considerably across OECD countries: The US doubled PC density in the 1990s and established a firm lead over EU countries, whereas Italy is far behind Finland, Sweden, Germany and France (see Tab. A8).

Computers are useful tools given adequate software to run them. Indeed computers and software are largely complementary. An interesting corrolary is the following: If goods i (computers/chips) and j (software) are complementary goods, a fall in the price of i will raise the price of j. The strong fall of computer/chip prices due to static and dynamic scale economies should raise the price of software in the long run, however, copyright protection of software was rather imperfect until the late 1990s – only then did US courts acknowledge that software could be patented. The software market certainly is a market in which intellectual property rights are very difficult to enforce. Given the imperfections of markets for knowledge in general and of software in particular, it is obvious that software is underpriced, and hence value-added in the software industry is not adequately reflected in the system of national accounting which relies on official prices and quantities shipped. Compared to an economic system with no imperfections in knowledge, the market price of software is too low and that of computers is too high. To put it differently, if software could fetch adequate prices in the marketplace the demand for computers would be lower. The imperfection of the software market has an important consequence to the extent that building computer networks has positive knowledge spillover effects: it amounts to an implicit subsidization of computer networks. If, however, one could not find spillover effects in econometric analysis (that is the production of computers is characterized by high productivity growth while using computers is not associated with positive spillover effects), there would be an inefficient overproduction of computers worldwide. As regards the productive potential of computers, one should be aware of the „generation problem“: With every new generation of new computers which enters the market after 3-4 years, millions of old PCs, say in the insurance companies, are replaced with more powerful models, although „static reinvestment“ in the sense of simply replacing the existing model by the same model after some years would have been fully adequate (however, the old model no longer is produced). To exploit the full potential of the new PC generation the firm often would have to qualify personnel which costs time and money. Such investment in human capital is relatively rare. The implication is that simple hedonic price indices for computers overestimate the real stock of computer capital.

Tab. A8: Computer Density in OECD Countries

	Personal Computers per 1.000 people 1998	Internet Hosts per 10.000 people July 1999
Austria	233,4	252,01
Belgium	286,0	266,90
Denmark	377,4	540,30
Finland	349,2	1.116,78
France	207,8	110,64
Germany	304,7	173,96
Greece	51,9	59,57
Ireland	271,7	156,68
Italy	173,4	68,28
Netherlands	317,6	403,49
Portugal	81,3	59,40
Spain	144,8	76,75
Sweden	361,4	581,47
United Kingdom	263,0	270,60
Japan	237,2	163,75
USA	458,6	1.508,77

Source: World Development Indicators.

The 1980s brought the mass production of computers. PCs were increasingly used by small and medium sized firms and even in households. New firms challenged established computer producers, including IBM, which had dominated the computer business over decades. IBM was a vertically integrated company which produced chips and computers, developed the necessary software and offered broad after-sales service. When IBM decided to outsource the development of a new operating system to Billl Gates, a new organization of the industry was born. Many PC producers used the Microsoft system from Gates, whose Microsoft Company become the most valuable firm in the 20th century. Intel produced chips (the engine of the PC business) which were used by several computer producers, but Intel itself faced strong competition. It then moved to the production of superfast chips for servers. The strong innovation dynamics in the electronics sector was equivalent to enormous technological progress as data storage, data processing and data transmission became cheaper and more reliable over time. Even more important, with modern PCs firms could organize production and marketing of products via large computer networks. The internet, which started only in the mid-1990s, opened up new opportunities for networking and also allowed a more internationalized and flexible way of networking. Open systems started to become popular in the late 1990s when Linux, developed by users all over the world, appeared to be a serious challenge for dominant Microsoft.

Leading producers of PCs in the 1980s and 1990s were the US, Japan, Korea, Taiwan, Germany and France. At the beginning of the 21st century the US had a clear leadership in the use of computers: 51% of the population, which is almost twice the EU figure of 28%. The continental EU countries face the problem of combining a low computer density with high prices for the use of the internet. Italy, given its low internet prices partly is an exception, but this is rather irrelevant since computer density in Italy is relatively low. There is a large and rather stable transatlantic computer gap to which corresponds an internet gap: continental EU countries are far behind the US, which seems to be defending its leadership. This is not surprising since PCs are the most important way for accessing the internet. Cable TV systems could become important in many EU countries, including Benelux, Germany, France and the UK.

Fig. A8: Computer Density in Selected OECD Countries, 2000 (households)

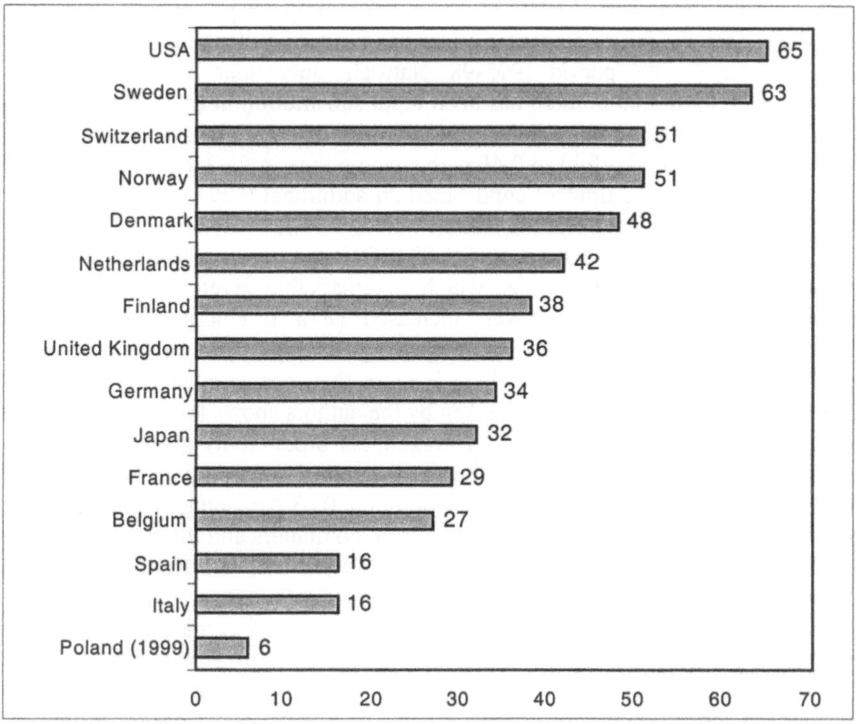

Source: *EITO/BITKOM (2000)*

The use of computers is one core element in the new digital economy. PCs are only one variant of computers. Advanced mobile UMTS telephones will match the first generation of PCs in almost every respect. It is unclear to what extent mobile telephony is a good substitute for PCs. If it is, Italy and the Scandinavian countries, who are world leaders in mobile telephony, should be the economic winners of the early 21st century. However, households computer density is very low in Italy and Spain.

Computers increasingly play a role for the internet, although less so for end users who are constantly being offered new ways of accessing the internet via mobile phones, "palm-type" electronic products, play stations and TV. Reliable servers for the internet supply side have become increasingly important at the beginning of the 21st century. In addition, computers naturally play a role for internet access in small and large companies.

JORGENSEN/STIROH (2000) empirically analyzed the possibility of an increase in total factor productivity growth in the US in the 1990s and how this can be explained, while SCHREYER (2000) presented a similar study for the G-7 countries. He finds that ICT capital goods crucially contributed to economic growth for all G-7 countries, with the USA in the lead. SCHREYER carefully takes into account the problem of price deflators for computers as they exist across OECD countries. Computer inputs contributed less than 1/10 of one percent to US output growth in the period 1959-73. Since 1973 computer prices have fallen fast, and since 1995 the fall of computer prices has further accelerated reaching almost 28% per year in the period 1995-98. Naturally investment in computers has strongly increased, that is there has been input substitution in most sectors and in particular in high-technology fields; contributions of computers to output growth increased more than fivefold to 0.46 percentage points in the late 1990s, software and communication equipment contributed an additional 0.29 points per year for 1995-98 (JORGENSEN/STIROH, 2000).

A major finding in the JORGENSEN-STIROH study is an acceleration of labor productivity growth in the 1990s, which is remarkable given the previous two decades of slow-down. Moreover, increased labor productivity growth was achieved after a decade-long economic upswing. Average labor productivity grew by 2.4% per year for 1995-98, which is more than one point above the period 1990-95. The decomposition undertaken by the authors shows that capital deepening added 0.49 percentage points to average labor productivity growth. Total factor productivity (TFP) growth (based on a production function with Hicks-neutral technological progress) contributed an additional 0.63 percentage points, "largely reflecting technical change in the production of computers and the resulting acceleration in the price decline of computers" (JORGENSEN/STIROH, 2000, p.5). Degrading labor quality growth has retarded growth of average labor productivity by 0.12 percentage points (relative to the early 1990s). This mainly reflected the fact that the pool of available workers became increasingly exhausted, given the US economy's approach to full employment. The JORGENSEN-STIROH study emphasizes the problem that computer-using industries like finance, insurance, real estate, and services have continued to lag in productivity growth. It seems that there have been few positive spillover effects from production of information technology to the sectors which use this technology, so there is weak empirical evidence for a new economy picture of spillovers cascading from IT producers onto users of this technology. This, however, overlooks the high growth of labor productivity in US telecommunications, and is also likely to underestimate the role of accelerated diffusion in the production function - the use of telecommunications has significantly contributed to economic growth in Germany in the 1960s, 1970s and 1980s (JUNGMITTAG/WELFENS, 1996; 1998, 2002). In a production function with capital, labor, knowledge (approximated by patents and real expenditures for imports of licences), and telecommunications the latter showed a significant impact on growth, although with declining contribution over time; a declining impact over time might be due to the effect that at low penetration rates the diffusion effect of using telecommunications is marginally higher than when high telecommunication penetration rates have been achieved.

Relatively deregulated markets in the USA and Scandinavia, as well as the respective structure of the economy, can largely explain the leading role of the US and Scandinavian countries in ICT expenditures. The UK figures have an upward bias because of the London financial center. While long distance and international competition in the US and the UK was already introduced in 1984, the EU countries (except for the UK) introduced competition only after 1998. Only some EU countries introduced competition in local telephony in 1998, but the European Commission and the European Parliament adopted a directive which will bring about competition in local telephony in all EU countries in 2001-2003.

2.4 Information and Communication Technology, Trade and FDI

It certainly is much too early to expect the internet to have contributed strongly to acceleration of economic growth, since the productivity-enhancing effects go along with adjustment costs associated with the new information and communication regime. Adjustment costs mean that employees and workers have to learn the new technologies so that a kind of fixed setup costs counterbalance in the early expansion stage of the internet the gross productivity effects of e-commerce, internet-based remote services, etc.

Fig. A9: Magic Triangle of Trade, FDI and ICT

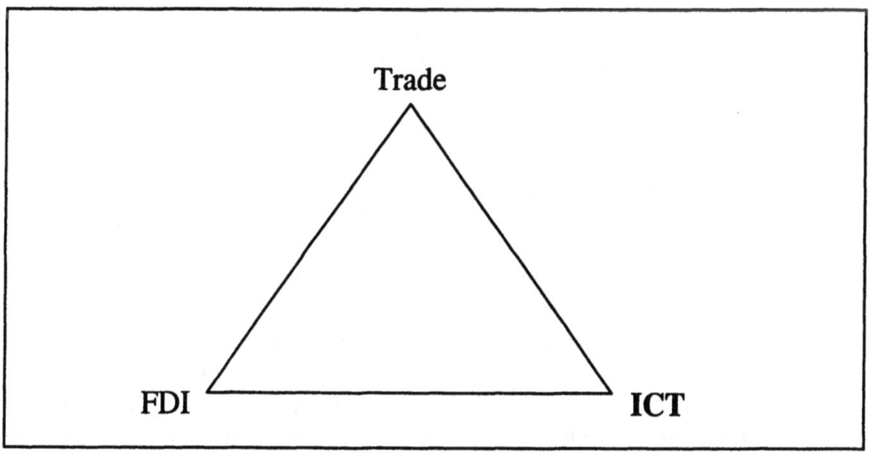

While empirical effects of the internet will be difficult to assess for quite some time theoretical analysis can be useful. In the following paragraphs we will look at the links between ICT, trade and foreign direct investment on the one hand, and the impact on economic growth on the other. One might indeed consider the interaction of ICT, trade and FDI as a magic triangle which stimulates growth; countries which are either weak in ICT or in FDI inflows are likely to fall back, while the leading OECD countries are strong in all three fields. Some arguments supporting such a triangular view will be presented in the following paragraphs.

ICT Dynamics

Information and communication technology refers mainly to computers, telecommunications, and the internet. The speed of transmission within the telecommunications sector is particularly important for the economic value of networks. This speed has drastically increased in the 1990s and is expected to further increase in the early 21st century as broadband networks (cable TV and xDSL technology in the existing fixed line telecoms network) is becoming more widespread. According to figures projected by VisionConsult in September 2000, the US will have 35 million broadband users by 2005, Europe 30 million, and the rest of the world some 20 million. These figures clearly suggest that the ICT revolution in the field of broadband network development mainly concerns OECD countries. Broadband internet access is particularly useful if one wants to mail or download complex images and videos. This should be standard in the business community by 2010.

The US had 2.5 and 5.5 million broadband users in 1999 and 2000, respectively, the percentage of internet users with broadband access reached about 7% in 2000. This is expected to increase to some 20% by 2003. This would imply that roughly 15% of the population would be broadband internet users if 75% of the population is online by 2005. This is still a small group of users who could easily benefit from the streaming video market. According to Usdata.com, the interactive video market will reach $4.2 bill. by 2005. Streaming videos generated some 170 million streams in 1999. This will grow to be about 1 billion streams by 2005. Cable TV provides up to 30mbps downstream and 10 mbps upstream, although a realistic transmission speed according to streamingmedia.com, is up to 1.5 mpbs downstream and 300kbps upstream. ADSL has a theoretical transmission rate of up to 8 mbps downstream and 640 kps upstream, with a realistic transmission rate of 1.5 mbps downstream and 32 kbps to 1.1 mbps upstream. SDLS has up to 1.5 mbps in both directions, while it's realistic transmission rate up to 770 kbps in both directions. HDLS has a theoretical and realistic transmission speed of up to 1.5 mbps in both directions. Compare this to ISDN which has 64 kb/s. With ISDN it would take 14 minutes to download a short videoclip of 7MB, while with DSL at 770 kbps it will only take 70 seconds.

The internet helps create truly global markets, and at the same time encourages the building of expansive international networks. There are only half a dozen of prime backbone network operators for the internet and those as well as smaller network operators are investing in growing international networks. A crucial element of the growing internet traffic is underwater fibre optic cables. In 1998 eleven major companies, including AT&T, BT, Cable & Wireless, Deutsche Telekom AG, France Telecom, KPN, MCI, PGE, Sprint, Swisscom and Telia, agreed to lay a new transatlantic telephone cable, the TAT-14 which will allow up to 14 million transatlantic telephone calls at the same time. In 1995 70% of underwater cables were used for telephone calls, but due to the internet and the rapidly growing data traffic this share will have fallen to about 15% by 2005 – data transmission will account for 85% of underwater cable capacity use. Technological progress in underwater capacity is enormous since the TAT 14 cable has 640 Gbits/s transmission speed (for four pairs of fibre optical lines) and 64 times the capacity of the TAT12/TAT13 cable. The new transpacific Japan-USA cable network, operational in 2000, is almost as powerful as the TAT14 cable with a transmission speed of 80 Gbits/s and a capacity of almost 1 million simultaneous transpacific telephone calls. Within a few years 680 Gbits/s should even be possible (KRISTIAN, 2000). While almost all existing transatlantic and transpacific cables are co-owned by leading national operators, there will be a few independent operators in the future (ie Global Crossing). However, one may doubt that independent operators could play more than a marginal role in transatlantic and transpacific cable markets.

Given the enormous increase in transmission speed, falling prices of telecommunications and the growth of mobile telecommunications, it will become increasingly easy to access new information and to communicate with people worldwide. This should reduce transaction costs in markets.

As is obvious from the above figures, there is enormous growth of transmission capacity both across the Pacific and across the Atlantic, but there will also be an enormous increase of data traffic. All this data has to be stored, at least temporarily, and this will enormously stimulate the expansion of large servers and of data storage systems, with IBM, SUN, and EMC2 leading in this field.

Trade and ICT

The internet is reducing transaction costs, which will stimulate international trade. Transaction costs drive a wedge between the gross demand curve (DD_0) and the net demand curve (DD^{net}). The intersection of the net demand curve and the supply curve determines equilibrium: In a closed economy the pecuniary price is p^E, representing unit revenue for producers, while the gross price is $p^{E'}$, reflecting the net price plus the information/transaction costs which are equivalent to the distance EE′. If we assume a small open economy facing a given world market price p^*_0 with transaction costs for imports equal to those at home, then we have domestic output at q_2 and point B is the quantity demanded. Hence imports equal the distance AB. Transaction costs are equal to the distance BB′. If the introduction of the internet would reduce transaction costs to zero, the gross demand curve would become the relevant demand curve (gross and net demand coincide), and imports would increase considerably. The new import would be equivalent to the distance AC. This partial equilibrium analysis ignores the positive income effect which is observed in general equilibrium analysis. Taking into account a positive real income effect implies that the gross demand curve would shift to the right. Furthermore, if the reduction/elimination of transactions costs has a pro-competitive effect in world markets, the world supply curve would shift downwards so that p^*_1 would hold and C" would mark the relevant intersection point for demand.

Fig. A10: Transaction Costs and Trade

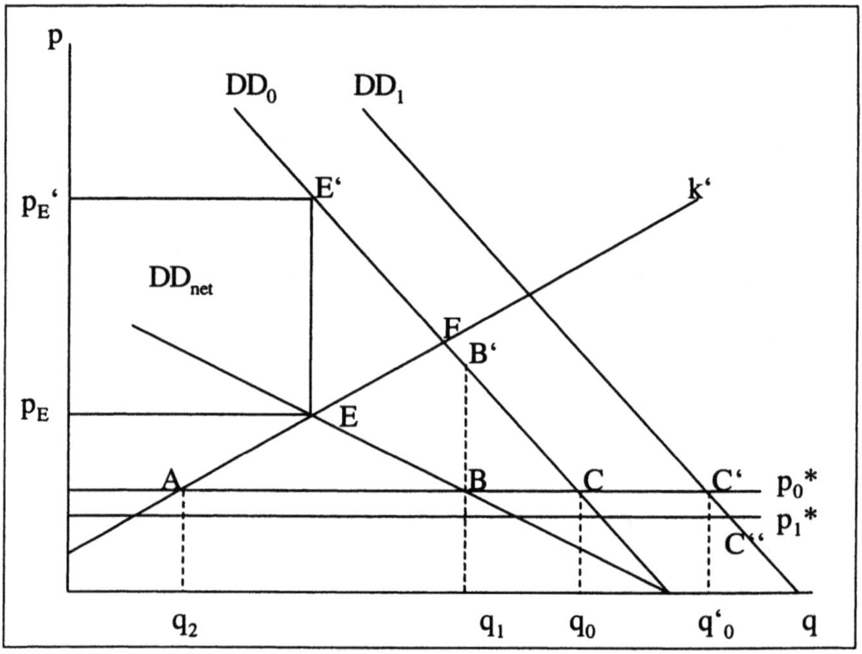

An interesting problem of international trade is those sectors where telecommunications is an important input for service j, where the input i is supplied by newcomers and the former monopoly network operator. Units have been normalized in such a way that one unit of i is required to produce one unit of j. Assume that all newcomers together are represented by the marginal costs curve k'_i, while the dominant operator, with much excess network capacity, is producing at constant marginal costs $k'_i{}^*=p^*_i$. We assume that value-added services are produced at marginal costs of k'_j in a competitive market which is facing an infinitely elastic supply from the world market at P^*_j. What happens if market demand in the j-sector in the country is DD_j while the dominant network operator enjoys some power in market i and decides to fix its offer price at $p^*_i\,(1+t'_i)$; alternatively, the regulator might have set the i-price at $p^*_i\,(1+t'_i.)$. The following diagram provides a graphical analysis of the problem. Note that the effective supply curve of the i-sector is a kinked curve AIT′ in the absence of market power of the incumbant operator, but AJS′ in the case of market power, which is reflected here in a top up on marginal cost. In a closed economy let us assume that the dominant operator has constant marginal costs in providing j so that a cost-based supply curve is given by p_1. Assuming that for the newcomers the marginal cost curve for value-adding in sector j is given by k'_j. The effective supply curve of the j-industry is given by a kinked supply curve CDE´ under full competition in the i-industry, and by CD´F´ in case that the network operator has market power in market i. Without market power in activity i the allocation would be as follows: demand is according to point N so that the j-quantity demanded is OH, the incumbant operator will produce j-services according to the distance XN (corresponding to GH), while OG is supplied by the newcomers. With the incumbant operator enjoying market power in sector i output of the newcomers in the j-industry will be smaller, namely OY´. As demand is unchanged the supply of the incumbant operator will increase: it will produce Y´H. If the incumbant operator also enjoys market power in the j-market so that instead of p_i the higher $p_i(1+t'_j)$ line, with t'_j denoting the quasi-monopolistic top-up, is his relevant supply line, the situation is slightly different. Demand will fall as characterized by point N´, the incumbant operator would produce j-services according to G´H´, and OG´ would be supplied by the new firms. This is a pardoxical result since the newcomers produce more j-output than under full competition; with the introduction of competition only in the j-sector the incumbant operator will increase its output. Those familiar with the theory of effective protection will notice that there is some similarity between this theory and the case of market power in a two sector setting with intermediate products (CORDEN, 1968).

Fig. A11: Market with "Monopolistic Imputs" in Partial Equilibrium Model

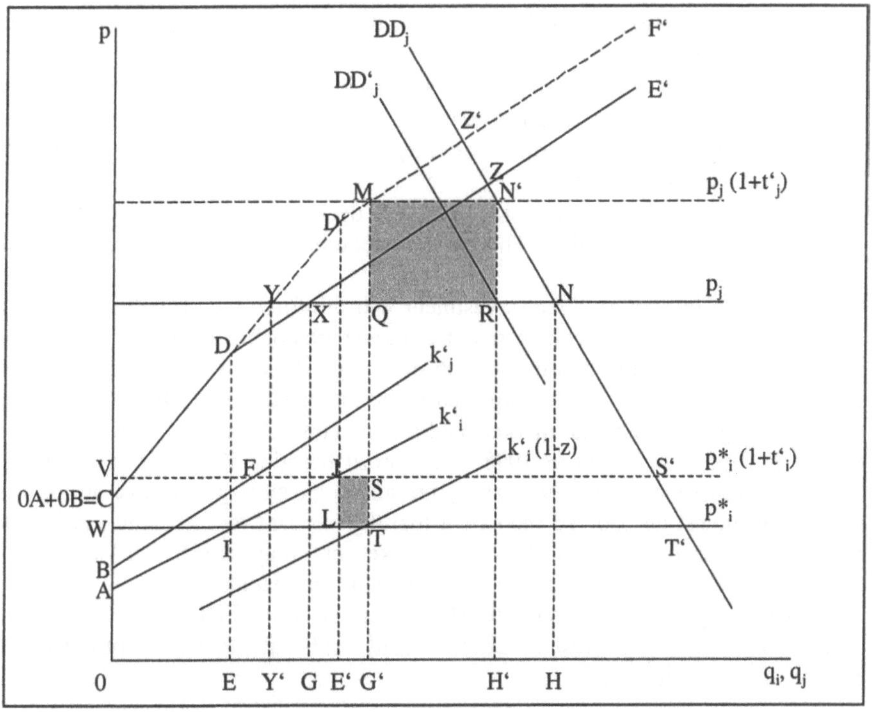

The internet enhances the international tradability of services. While services account for about 60-70% of value-added in output in leading OECD countries at the beginning of the 21st century, services in overall trade of OECD countries accounted for only 20%. The introduction of the internet is likely to increase the share of traded services considerably. Since trade is known to stimulate productivity growth the trade-enhancing effects of the internet could indirectly contribute to higher economic growth.

The 1996 WTO liberalization of telecommunications will stimulate sustained growth of trade in telecommunications services, which have been restricted for decades. Since there still is no full liberalization the WTO should keep the topic on the agenda.

Since transaction costs in markets and intra-firm transaction costs will not be reduced equally across sectors, one may well observe, following a standard COASIAN approach, that market transactions will increase in those sectors where the relative decrease of market transaction costs are relatively high to savings of firm-internal transaction costs.

ICT and FDI

Modern computer and information technology should stimulate foreign direct investment for three reasons:

- ICT particularly supports the expansion of knowledge intensive industries. Hence there will be more firms with ownership specific advantages which will seek to exploit such advantages through international foreign direct investment.
- ICT is itself a rather innovation intensive sector (broadly defined). This means that foreign direct investment in the ICT sector itself should grow strongly. Vertical international integration of firms can be expected to grow strongly.
- ICT allows firms to organize in a more decentralized fashion across a larger geographical radius. This should encourage expansion of multinational companies.

The growth of foreign direct investment will reinforce the growth of trade, to the extent that foreign direct investment and trade are positively correlated.

One should also note that the growth of foreign direct investment in the telecommunications sector can be expected to be rather strong in OECD countries, since opening up of national markets and privatization of national network operators in Europe only began in 1998. With profit margins shrinking in gradually more competitive markets, the (partly) privatized ex-monopolist have strong incentives to invest abroad. To the extent that the relevant market is no longer the national market, but rather an EU-wide market, one may expect that leading national network operators will try to build EU wide companies.

At the bottom line one should clearly expect that trade, FDI and ICT will reinforce each other. This virtuous circle should contribute to higher productivity growth and GDP growth.

The analysis conducted so far has shown that trade, FDI and CTI go together. In Economics there is ample evidence that trade and FDI contribute to economic growth. The more complex question is, however, how channels of international interdependency can be taken into account in the production function.

2.5 ICT Dynamics and Entrepreneurship

The modernization of telecommunications, in combination with falling prices of telecommunication services and computers reduces barriers to entry for newcomers in many fields, in particular in the service sector. There is no doubt that the US has been much more successful in creating new digital firms than the EU; and of more importance, the number of digital firms which obtained a listing in NASDAQ is much higher than in Europe's new stock markets.

In the US, digital entrepreneurship is often based on successful spin-offs from university activities. The US culture of entrepreneurship encourages newcomers in the New Economy. The situation is different in most EU countries where the tradition within universities emphasizes to a far lesser degree application-oriented research. Incentives in relation to university-based entrepreneurship are also very weak. In addition, it is noteworthy that it is typical for websites of leading US universities to proudly mention that for instance x firms have been created by alumni. In contrast, very few German, French or British universities display entrepreneurial activities of alumni. The situation is slightly better in Sweden and the Netherlands where pro-business attitudes gained ground in the 1980s and 1990s. Among the smaller countries in the EU, Ireland shows a successful example of combining indigenous entrepreneurship and multinational investment, stimulated in part by government strategies. The government has, in the past 20 years, particularly emphasized education and retraining on the one hand, on the other, Enterprise Ireland (an agency promoting SMEs and MNCs willing to invest in Ireland) has nurtured domestic entrepreneurship by investing limited capital into new firms; while government had a share of more than 25% in the 1970s, it wisely decided subsequently to take a much smaller share. This was in order to avoid being forced to pay in additional capital, in the case of the firm being unsuccessful, naturally a common problem due to the high failure rates in this sector. With a minority share of 10-15% there is obviously a clear positive signal to other investors and a strengthening of the company envisaged, at the same time government's threat to pull back is much more credible than if government agencies took rather high stakes in start-up companies.

A look at Asia is particularly interesting in the IT sector. Many firms, both large and small, expanded into the IT sector in several Asian NICs in the 1990s. This expansion has been supported by government expenditures in the field of R&D promotion and by efforts to improve the level of education which is crucial for the skill-intensive IT sector. Taking a look at selected Asian NICs (see table), the share of overall exports in the fields of computer, telecommunication and IT components going to OECD countries represented between 20 and 77% of all exports (we disregard Indonesia here). As long as there was high IT investment in the US – and Europe – Asian NICs benefited from high growth rates – see the figures for the first half year of 2000. However, as IT investment is rather cyclical, the Asian NICs will also have to cope with cyclical downturns in the US and Europe, a problem in 2001.

Tab. A9: High Technology Exports of Asian NICs

	Share of Total Exports of Economy Going to OECD									US-Imports: Change (%) (during year until)	
	CN	HK	ID	KR	MY	PH	SG	TH	TW	June 00	Feb. 01
Computer[1]	6	7	2	13	19	22	54	16	28	8	–6
Telekommunication[2]	7	4	5	6	15	6	5	7	4	43	1
IT Components[3]	8	18	2	23	24	33	17	11	17	22	–2
Total	20	30	9	41	58	60	77	34	50	19	0

Source: BIS 71st Annual Report, Basle.

2.6 Internet and Company Dynamics

Telecommunication-based and internet related services could play an enormous role among OECD countries in the future. Expected sales figures for e-commerce show rapid growth in the US and the EU for the period 1995 – 2005. The more important changes could, however, concern the organization of firms and the dynamics of markets. Modern ICT allows large companies to be organized in a more decentralized fashion. Moreover, transaction costs within firms can be reduced via the internet; therefore large companies will continue to play an important role in the digital economy. It is true that the internet also reduces transaction costs in markets, but the more important aspect is that the internet will accelerate knowledge diffusion and thereby stimulate innovation. R&D-GDP ratios increased in the 1990s, Sweden achieving the top ratio of 4%, roughly double the figure of the 1960s. The end of the Cold War has reinforced global Schumpeterian dynamics in civilian markets for tradable goods and services as the US, France and the UK have raised their respective share of civilian R&D in overall R&D budgets. Based on the OLI-approach of DUNNING we may therefore assume that the role of multinational companies will grow: ownership specific advantages will play an increasingly important role in the knowledge society which will strongly build on ICT and the exploitation of network effects. This implies that entrepreneurship in the digital age will face at least four new challenges: Entrepreneurship will have to be

- Increasingly based on knowledge and technology
- Increasingly internationally oriented after a relatively short domestic expansion stage
- Increasingly oriented towards exploiting network effects
- Increasingly prepared to deal with large companies which will rely on inputs of innovative SMEs.

For firms, there exist internet-based opportunities concerned with intra-company communication, however, most internet-based opportunities will have a focus on flexible and differentiated production, on the flexible sourcing of inputs or complex after sale services. R&D activities could also benefit from the internet, however, as tacit knowledge will play a crucial role, personnel communication and regional networking will continue to be of great significance in the digital age. The internet is nevertheless quite useful as an organizing tool for international research groups, it thus remains to be seen how the internet will affect R&D productivity. For SMEs the internet has created new opportunities to get into contact with highly specialized firms and research groups which would be otherwise quite difficult to locate. Powerful search machines are useful for innovative SMEs.

Tab. A10: Opportunities for Firms in the Internet

Production	*Personnel Recruitment*	*R&D Activities*	*Sourcing of Inputs/After-Sale Service*
Intra-Firm Networks	Expert Fora (e.g. Euro group)	Training and Joint R&D Activities	Cooperation with Subsidiaries
Market Perspective (e-sa-les/built-to-order-system)	Marketing (Internet as info platform/ branding asset)	Public Relations: Industry, Investors, Government	After Sales Service

Modern telecommunication networks now make it possible – due to digitization and fast data traffic – to provide increasingly differentiated services, especially via Internet. While there are still many questions related to the security of payments on the web, there is no doubt that the potential for growing electronic commerce is enormous. The OECD ministerial conference "A Borderless World: Realizing the Potential of Global Electronic Commerce" (Ottawa, 7-9 October 1998) came to the conclusion that (OECD, 1998a, pp. 3f):

"Electronic commerce offers a radically new way conducting commercial transactions, and is potentially a key engine to increase economic growth, and enhance development around the world; cooperation amongst all players (government, consumers, business, labor, and public institutions), as well as a social dialogue, must be encouraged in policy making to facilitate the development of global electronic commerce in all countries and international fora, and that their actions should strive to be internationally compatible whenever possible; governments should promote a pro-competitive environment to allow electronic commerce to flourish, work to reduce and eliminate unnecessary barriers to trade, and act where necessary to ensure adequate protection of key public interest objectives in the digital world just as they do in the physical world..."

From a long-term perspective, the global internet will lead to a global information age with considerable structural shifts and new trading and investment opportunities (CARNOY ET AL. 1993). Countries that create stimulating regulatory and other political frameworks for the expansion of the internet and of internet services could achieve strong income and employment growths. This makes adequate deregulation of the telecommunication sector a crucial necessity for the unfolding of ICT opportunities.

An important aspect of modern information and communication investment is that investment in computers and new software should considerably raise telecommunication productivity growth. Conventional output and productivity measurements underestimate, to some extent, the growth effects of computers and telecommunications (SCHREYER, 1998). There exist problems with the measurement of real output in information, computer and telecommunication-producing industries and the share of these industries of overall output, which is increasing over time. The share of the G-7 countries in manufacturing value-added increased from 8.7% in 1980 to 11.6% in 1995. Quality adjustments of final expenditure measures show that the share of expenditure on information and communication products has not stagnated over time but increased in selected OECD countries. The increasing use of information and communication technologies in work, considerably encourages labor productivity growth (OECD, 1998a). It can be further noted that the use of the internet does stimulate the dissemination of knowledge.

Until the beginning of the 1990s the majority of information and communication systems were implemented only in internationally active businesses. This was for several reasons including a low level of development and high purchase costs. However, this condition changed rapidly due to the emergence of the World Wide Web (WWW) as well as the growth of major internet service providers. Since this development generated cheap and easy access to global information and communication networks, self-employed and SMEs, could also afford to utilize these new opportunities and integrate new ICT into their business routines (SCHMID/ ZIMMERMANN, 1997). Moreover, the number of new firms in the telecommunications business has strongly increased after 1998 in Europe.

The use of ICTs depends considerably on the structure of the enterprise. Above all, large-scale enterprises use them to adapt internal organization structures to meet volatile market demand and to thereby support flexibility and modularization of hierarchical business structures. This also enlarges the external flexibility of these companies and enables them to efficiently expand in markets with differentiated products.

A major advantage of ICTs lies in the opportunity for quick access to information and communication with external partners such as business associations, chambers of commerce, public authorities (e.g. local or federal governments), customers and subcontractors and any other information providers accessible via the net. Significant structural disadvantages of SMEs lie in the limitation of material, personnel and financial resources (EUROPEAN COMMISSION, 1994). SMEs can gain particularly by flexible cooperation. Such cooperation is facilitated by modern information and communications technologies. The development in particular of the internet and its services have created standards that can also be used for the creation of regional networks. These in turn could be the basis of dynamic clusters.

The European Commission has launched several initiatives which are supposed to stimulate SMEs and regional development. In 1994 the EUROPEAN COMMISSION (1994a) launched the "Interregional Information Society Initiative" (IRISI). The objective of this small initiative, which was financed under the innovative measures of the structural funds, was to raise awareness in regards to the information society between people, regional stakeholders and the private sector and to develop actions in the framework of the information society which are likely to influence the regional economic structure. A successful program "Regional Information Society Initiative" (RISI) was initiated by the Commission in 1995. Structural funds in the period 2000-2006 are partly supposed to stimulate regional ICT development. Regarding interventions supporting the information society, the Commission stresses in its guidelines for the programming of these new measures, the ICT's role as an engine for regional development and cohesion (EUROPÄISCHE KOMMISSION, 1999). Precondition for the access to the information society - and hence for exploiting the opportunities coinciding with modern ICT - is the endowment of the regions with a sufficient telecommunications infrastructure.

Even in less developed regions, according to the Commission, investments in telecommunications infrastructures can be seen as profitable, meaning that private investors will make the necessary financial engagements (EUROPÄISCHE KOMMISSION, 1999, p. 5). Furthermore, the Commission argues that the interventions in 2000-2006 should focus on the stimulation of new services and innovative applications as well as on the provision of know-how for potential users. Keeping in mind that leading regions (e.g. the London area) can provide free access to the internet, whereas in cohesion regions the costs for the installation of a telephone main line alone, are about one third higher than in the densely populated core European regions.

The European Commission has stimulated the growth of trans-European networks and has encouraged web-based cooperation of firms in the EU. In the light of EU eastern enlargement, it will be only natural to extend the existing programs to eastern European accession countries. This in turn will create new opportunities for entrepreneurs in eastern Europe in order to conquer markets in western Europe and elsewhere. As regards the internet it is, however, unclear whether European firms will become a true challenge for US firms, which dominate the supply side of the internet. Roughly 50% of all servers are from the US and even in the UK a majority of servers were run by US companies in 2000. If the internet should turn out to be a true gateway technology, the full unfolding of the network potential of the internet will take some time, implying that at the end of the 1990s only some of the basic productivity effects had become visible; critical lags in network effects are at least suggested by DAVID/BUNN (1988) in their discussion of the electricity supply industry.

Countries whose innovative firms strongly benefit from the internet and ICT, will record at first a strong rise in patents, followed by a rise in net exports in the respective fields. This is a lesson which one can draw from the empirical analysis of GRUPP/MÜNT/SCHMOCH (1994). The acceleration of the US patent applications in the second half of the 1990s thus suggests that the US current account deficit should – ceteris paribus – decline in the future. However, modern ICT has also accelerated the product cycle. The relocation of technology intensive products to newly industrializing countries is taking place faster than previously. Asian

NICs have specialized strongly in IT products. They stand to benefit from future relocation of production from the US to Asia. At the same time they will become increasingly vulnerable to cyclical developments in the US and Europe, which are their main export markets for advanced IT products.

2.7 Competition in and Deregulation of Telecommunications

At first sight it is obvious that internet dynamics and modern telecommunication networks are linked with each other, but taking a closer look at the problems shows that one can have rather different approaches in regulating telecommunications and the internet. The following analysis looks mainly at US and EU approaches. The internet has traditionally not been regulated in the US, where government left the initiative for developing new applications and user groups to the private sector; this has resulted in global leadership in the internet society, except for mobile internet applications. Governments in EU countries have been more active in developing programs for the internet society, although the regulatory problems and approaches in the EU are different from those of the US.

Looking back on the 1980's and 1990's, one can see different historical points of departure (LAFFONT/TIROLE, 2000; PELZEL, 2000; SCHWARTZ, 1997). Until 1984 AT&T ("Bell System or "Ma Bell") was a vertically integrated monopoly which offered long distance service and also owned local operating companies (BOCs, later known as RBOCs). There were also a few independent local operators, also enjoying a local monopoly, with a market share of about 20%. As a consequence of an anti-trust case, AT&T ("Ma Bell"), which had a market share of close to 90% in 1980, was split up into several RBOCs (Regional Bell Operating Companies or "Baby Bells"), which had a monopoly at the local and regional level. MCI had obtained approval from the Federal Communication Commission to offer microwave communications between St. Louis and Chicago in 1969. This development, and MCI´s entrance into the market for switched long distance service, naturally brought a debate about interconnection with the local operating companies of the Bell system. It was the Department of Justice which filed an antitrust suit against AT&T in 1974; the department perceived that AT&T used bottleneck facilities, namely its local phone network, in order to favor long distance operations to the disadvantage of other interexchange carriers (ie Sprint, MCI). A consent decree, the "Modification of Final Judgement", led to the 1984 breakup of the Bell system. AT&T kept long distance operations, the R&D facilities (Bell Labs) and its manufacturing subsidiary (Western Electric). The new AT&T did not face restrictions with respect to lines of business. However, the seven divested RBOCs were restricted to local telephone service. The RBOCs, GTE (large non-Bell operator) and other smaller companies are called incumbent local exchange carriers (ILEC). All regions served by the Bell system were divided into local access and transport areas (192 LATAs). All interLATA service was considered as long-distance service so that it was covered by long distance companies.

The RBOC's desire to lift the restriction to local service was successful in 1996 when the new Telecommunications Act was adopted. The basic idea of the Telecommunications Act is to allow the creation of companies which can offer a broad range of services including voice telephony and internet. The act aims at creating local competition and allows RBOCs to enter the long-distance market once there is adequate competition in local markets. A main aim, therefore, is to create competition between ILECs and competitive local exchange carriers (CLECs), where market entry can take place in three ways: facilities-based, via resale and via unbundling. The latter means that entrants can lease elements of the local network (ie using ILEC´s wires but provide its own switches). The FCC sets default/standard prices for unbundled elements to be applied by states that have been unable to conduct cost studies, which are typically based on long run incremental costs. The Telecommunications Act of 1996 and the corresponding FCC order consider three elements for interconnection, that is when an entrant must interconnect with an incumbant so that each provides termination of calls originating from the other´s network.

1. Nondiscrimination: The ILEC has to offer identical interconnection conditions to all telecommuncation carriers, so mobile operators and cable TV companies obtain the same conditions (enforcement is guided by a quasi-most-favored nation treatment). However, CLECs are not required to provide other operators the same rates that they use in reciprocal deals with ILECs.
2. Cost-based pricing: the incumbant firm´s charge has to be based on a cost study which is forward-looking (based on total element long-run incremental cost =TELERIC).
3. Symmetrical access charges: ILECs have to accept reciprocal compensation arrangements with other telecommunications carriers where the symmetry of access charges is mainly motivated to avoid the problem of the incumbant operator. States have the right to opt for a bill- and-keep-arrangement where termination charges cancel out on both sides, assuming that traffic between the networks considered is nearly equal and can be expected to remain so in the future.

The new AT&T maintained a strong position in long distance telecommunications where it faced MCI, Sprint, and other competitors. Under the 1996 US Telecommunications Act, market access was opened on a kind of reciprocal basis, that is regional baby bells were allowed to enter the long distance market provided that they opened up the local market for long distance operators. The Telecommunications Act states that a RBOC will be allowed to enter the long-distance market if (i) it has entered into at least one access and interconnection agreement with a competiting facilities-based carrier or at least offers such agreements under conditions approved by the state; (ii) it must comply with a checklist which includes various requirements related to unbundling, resale, interconnection etc.

In addition, cable telephone companies were also allowed to enter the telecommunications business; until 1996 US telecoms operators could only go abroad to test the new opportunities for combining cable TV services, telephone services and fast broadband internet services, and many entered the UK in the early 1990s with such services. There the government had adopted a duopoly approach in

1984 when British Telecom was privatized and long distance competition from Mercury (a subsidiary of Cabel & Wireless) was allowed.

In the US cable TV operators were facing stiff competition from satellite operators so that they quickly explored new business fields in the 1990s. After 1996 AT&T entered local telephony mainly with a newly acquired cable TV company which offers cable TV, internet services and telephony.

The situation in EU countries was rather different when the telecommunications markets opened up under a directive of the European Commission. Opening up concerned the network operation and network services. The main problem was that, except for Finland, national telecommunications operators had a full monopoly position in the sense that they controlled the access market (local telephony) and the long distance, as well as international telecommunications traffic. International telecommunications traffic was priced according to the rules of the International Telecommunications Union (ITU) which is an international organization based in Geneva. The basic principle of pricing was that country A (originating traffic) and country B (terminating calls) would apply a so-called accounting rate for the traffic volume which cancelled out. This does not exactly encourage efficiency given that accounting rates applied can be expected to be shaped by the less efficient companies and countries. The countries with a monopoly (which can be expected to be rather inefficient and charge high prices) will thus have a strong influence on the accounting rates.

Competition in telecommunication does not come automatically. For decades it has been argued that telecommunications is a natural monopoly as average and marginal costs fall when the network expands. This is the so-called natural monopoly problem. The traditional argument was that creating a competing second network would mean duplication of investment, so governments should accept a (private or public) monopoly network operator: governments accepted telecommunications monopoly but imposed price regulation and certain standards, which in itself has some costs. The extent of economies of scale in fixed-line telephony, based on analogue signal transmission, is an empirical question. Modern digital cellular telephony is obviously characterized less by traditional scale economies. In many EU countries several networks are competiting against each other profitably. A crucial element of digitization and digital telecommunications concerning both fixed and mobile network operation is that fibre optical lines can be split easily. Transmission capacity can be split so that the problem of excess capacity is not a serious one in modern telecommunications; competing network operators can lease capacity, and one may even anticipate a future market for transmission capacity, a market which is being developed by Enron and others in the US. To the extent that one can separate network operation and telecommunication services one might consider government regulation that prevents anticompetitive vertical integration in order to achieve competition at least in the field of telecommunication services. If, however, competition is both possible in network operation and in the provision of services this would be beneficial to society and consumers.

The analysis of networks has recently intensified (eg KATZ/SHAPIRO, 1994; ECONOMIDES, 1996, WELFENS, 1996). Crucial in the field of telecommunications is that with more users linked to the network it will become more valuable to each user. This positive demand side externality is external to the user because his decision to join a network is based only on the private benefits gained by him/her. This means that there could be some tendency of private investors to underestimate demand growth in the stage of network and services expansion. In markets with tight oligopolies one may, however, assume that each firm takes into account both the reaction of its rivals and the well-known network effects which will benefit all operators.

Given the natural monopoly problem it was obvious that regulation – price caps (a maximum growth rate n for a bundle of telephone services relative to the inflation rateπ, typically $n<\pi$) or rate of return regulation – would be adequate in order to avoid that the monopoly operator would exploit consumers and firms using telecommunication services: The economic benefits from falling marginal costs would be diluted if the company were allowed to impose monopoly prices. For decades EU countries have chosen government ownership and regulation by a ministry of telecommunications as a strategy to cope with the problems described. Only under the pressure of the European Commission did EU countries decide to open up fixed network operation and telecommunications services for competition and to introduce relatively politically independent regulatory authorities which imposed asymmetric regulation in voice telephony as a means to cope with the dominant operator – rules mainly concerned price caps, billing, number portability and unbundling of the local loop (the latter in some countries: Austria, Denmark, Germany, Finland, Italy, and the Netherlands). All EU countries quickly embraced gradual privatization of the ex-monopoly network operator. Incentive regulation, based mainly on price caps, was adopted in most EU countries after 1998. The price cap is typically based on a basket of services which leaves the telecommunications company considerable discretion in fulfilling the price cap requirements. While rate of return regulations are known to encourage capital intensive provision of services, price caps are expected to encourage efficient behavior of the firm since it is free to generate high profits and to exploit options for cutting prices as long as the price cap is met.

With the advent of digital cellular phone systems the natural monopoly argument became less convincing than before. Moreover, since cable TV companies could also in principle provide telecommunication services, there is indeed some competition in fixed network telephony as soon as there is cable TV; a satellite system is certainly a much weaker substitute for fixed network telecommunications.

When the US decided to split telecommunications in a local (cum intrastate) telephone monopoly system and competing long distance carriers in 1984, the basic idea was that the natural monopoly problem was mainly one of the local loop with the notorious so-called „dirty last mile", while competition in long distance and international telephony faced no such problem. The philosophy of the Telecommunications Act of 1996 was rather different since it opened up local markets for entry of long distance companies on a kind of reciprocal basis. A major problem which already occurred in 1984, namely rules for interconnection, came up in a modified form in context of the 1996 Act. The US has had rules on line sharing since 1999, that is a given telephone line can be split for voice services and DSL, which could be used for super fast internet services.

As regards its leading role in the New Economy development, the US benefits from strong competition in telecommunications following the dismemberment of AT&T in 1984 and the 1996 Telecommunications Act. In 1984 AT&T accepted the modified judgement requiring that the old dominating AT&T (often called „Ma Bell“) be split in a long distance carrier („new“ AT&T) and several regional „Baby Bell“ operators which would handle local and some interstate telephone calls. After a short transition there were seven Baby Bell companies. AT&T faced increasing competition in the long distance market where MCI and Sprint (after 1989 also WorldCom) became strong competitors over time: AT&T´s long distance market share fell from 90.1% in 1984 to 43.1% in 1998, while that of MCI and Sprint increased from 4.5% and 2.7% in 1984 to 25.6% and 10.5%, respectively, in 1998; WorldCo had about 7% in 1998, and all other long distance carriers raised their share from 2.6% to 20.9% in 1998. The Herfindahl-Hirschman index fell from 8.2 in 1984 to 2.6 in 1998, interstate switched access minutes increased strongly in volume in the period 1985-1998: from 167 billion minutes to 519 billion minutes. The 1996 Telecommunications Act has opened the local and regional loop for long distance operators, while regional operators are free to enter long distance telephony. This has contributed to a rising number of both long distance and local loop operators, including operators from cable TV business. Overall 1998 revenue in the telecommunications industry reached $246.4 bill.; $197,5 bill. was from end user revenue of which $74.8 bill. was end user revenue in local service, 33.5 bill. was for wireless service and $89.2 bill. for toll service; revenues from intrastate services and interstate services amounted to $122.5 and $74.9 bill., respectively. Carrier´s carrier revenue totaled $45.5 bill. in 1998, of which $29.2 bill. was for local service, $3 bill. for wireless service, and $13.3 bill. for toll service; intrastate and interstate service amounted to $18.8 and $26.7 bill. Carrier´s carrier revenue includes, among other items, access services that local exchange carriers provide to toll carriers (FCC, 2000).

In the US the internet is completely unregulated. Firms get access to the internet mostly via business calling plans of telecommunication operators which bundle various value-added services including an internet browser and access to the internet. The mass consumer market, however, is in the hands of specialized internet providers. Since there is a flat rate for local telephony in the US (except for a few regions), internet access implies no extra telecommunication costs for the residential user. However, private households have to pay the internet service provider. Taking the example of the leading US internet service provider (as of mid-2000), AOL, there is a monthly flat rate membership of $21.95; alternative offering of three hours is $4.95 per month, with additional time priced at $2.50 per hour. Another alternative offer is $9.95 per month for unlimited use for subcribers who have an internet connection other than trough AOL and use this connection to access AOL services. Consolidated revenues reached $6.9 bill. in the fiscal year 2000 (ending June 30, 2000), up 43% over the previous fiscal year, subscriber growth was 5.6 million, which brought the AOL world total to 23 million members by June 30 (AOL, 2000, p.3). Net income rose to $1 bill., while advertising, commerce and other revenues reached nearly $2 bill. for the fiscal year 2000.

The enormous success of AOL in the US (to the extent of taking over the old economy giant TIME WARNER) is surprising given that regional telecoms operators could have easily launched their own internet service plan. However, the regional telecommunications giants did not understand the internet business. They totally underestimated it, and when they finally noticed that it would become a big business it was to late to launch a branded ISP service successfully – first mover advantages obviously matter strongly in the internet world.

When the National Science Foundation entrusted private backbone operators to run its system there was the problem of interconnection. With several backbone companies expected to cooperate, NSF created in the early 1990s a system of geographically dispersed Network Access Points (NAPs) where each point consisted of a shared switch or local area network used to exchange traffic. In 1995 this network of commercial backbones and network access points permanently supplanted the NSFNET; the backbone companies developed a system of cooperation and interconnection, which is known as peering: Peering partners, say backbone A and B, exchange traffic that originates with the customer of one backbone and terminates with the user of the other peered backbone. Peering however does not mean that A will be an intermediary for C´s traffic destined for backbone B. As a result of congestion, the NAPs several backbone companies started to interconnect directly with each other, a system known as private peering (KENDE, 2000). Not all backbone firms have peering arrangements meaning mutual free exchange of data packages, the alternative is to pay for transit, where price is fixed in a transit arrangement. Negotiating adequate prices is difficult since one can hardly focus only on the traffic volume; geographical spread, capacity and the number of customers also matter. Taking into account the implicit rule of peering, what matters is that there is a comparable value in traffic exchanged.

While the US telecommunications regulator has not yet imposed any regulations for the internet, there are concerns (KENDE, 2000) about the emergence of a dominant backbone provider which could undermine competition between companies that own the internet´s infrastructure. However, KENDE argues that backbone companies governed only by ordinary laws of contract and property, and subject to antitrust rules, are likely to provide interconnection to other networks, including those offering innovative service; his paper shows how competition, shaped by antitrust laws and competition enforcement can prevent the emergence of a dominant firm. This acts to restrain the anti-competitive action of larger backbone companies in place of any industry-specific regulations, such as interconnection obligations. The FCC could oppose mergers that increase the market share of the leading backbone operators. A dominant backbone operator might have no real incentive to connect with smaller backbone providers which might trigger regulatory intervention of the FCC. Until the beginning of the 21st century strict antitrust enforcement helped to avoid any critical situation in the backbone market; ie MCI was forced to sell off its backbone before it was allowed to merge with WorldCom (also note the blocking of WorldCom´s acquisition of the long distance operator Sprint). Large backbone operators of similar size typically allow reciprocal free interconnection, but the relationship between large and small backbone operators is different. Small operators such as Exodus and Level3 have complained that leading providers (ie UUNet/subsidiary of WorldCom) are not offering free interconnection. Instead they would charge the small operator to pass off web traffic. The FCC obviously holds the view that transit charges for small operators are not inadequate – there are positive marginal costs if smaller operators are allowed to use the large operator´s infrastructure. In an international perspective, transit charges of large US operators are rather modest, although Australia and Asian countries have complained that the cost of transporting data across the ocean, as well as the transit charges of US backbone companies. Reflecting high telecommunication costs in Asia, and lack of competition in Asia, a considerable share of Asian internet traffic travels via the US; this development shows the powerful logic of international price arbitrage. The International Telecommunications Union has adopted (based upon proposal from Asian countries) a recommendation in spring 2000 that would lead to a cost sharing arrangement similar to the way voice circuits are priced. The US and the UK, as well as other countries, have opposed this recommendation.

With digitization of signals and fibre optical cables there are new opportunities for trade in transmission capacity. This means that the problem of overinvestment emphasized in the analogue fixed network is indeed less relevant because flexible pricing and splitting of capacity facilitates the problem of recovering investment costs for the network operator.

In the internet society cheap and fast access to the internet is crucial. Here telecommunication infrastructure plays a crucial role, and here there is not only a North-South divide but also a divide within the OECD countries. The US, Canada, Australia and New Zealand have flat rates in local telephony and since the internet access goes via local telephony these countries have the opportunity to jump-start the internet. For users it is easy to calculate what the internet access will cost, and the overall price of the internet is indeed rather cheap in these countries. Since using the internet means data transmission, reflecting voice, images or text, fast data transmission becomes important. Here new fibre optical cable networks and use of powerful servers becomes crucial. Building global networks represents new profit opportunities, but it also requires massive investment. These investment needs have already started to profoundly affect international financial markets as telecommunication companies and internet firms have become top borrowers in bond markets, while at the same time the internet itself is affecting global networks and raises new issues about cyber cash, financial engineering and data protection.

European Deregulation

When the European Union set 1998 to be the liberalization date for telecoms network operation and telecoms services, few analysts anticipated the dynamics unleashed by this policy move. Within three years continental EU countries with their formerly low-productivity national monopolies have switched to a very different setting. By mid-2000 all EU countries (Sweden with the national operator

Telia being the latest in January 2000) have partly or fully privatized the national network operator and opened up for national and international newcomers. Facing aggressive newcomers and parallel competition from mobile networks, national operators have sought to expand abroad and to form larger companies through mergers & acquisitions in this highly competitive and technologically dynamic market. Technological dynamics means that old copper cable networks are increasingly outdated, although DSL technologies can be employed to considerably increase the speed of transmission. New fibre optical cables are laid by aggressive newcomers such as Quest or Level 3 in the US, or Colt (City of London Telecom) in Europe. Facilities-based competition of newcomers, including foreign ex-monopoly operators, is one element of the new market dynamics in the EU. A second strand of competition is based on leased lines and the resale of capacity. Resale based competition cannot be expected to dominate market dynamics, but resellers are an important part of the game. They help to minimize excess capacity, which they on the basis of flexible pricing.

Facing asymmetric regulation in EU countries, with dominant ex-monopoly operators having to obey certain pricing rules while newcomers enjoy freedom of pricing, the newcomers have certain opportunities in the liberalized EU framework. However, to the extent that a pan-European market is emerging, pure resellers are likely to face problems as they are not creating pan-European reseller activities. Since national regulators have adopted rather different approaches in individual EU countries, creation of EU-wide reselling activities and strategies is rather difficult. The EU deregulation in telecommunications and in energy has indeed raised complex problems for regulating network industries (BERGMAN/ DOYLE/GUAL/HULTKRANTZ/NEVEN/RÖLLER/WAVERMAN, 1998).

Since the development of GSM mobile technology the use of telephone networks has a meaning. GSM is a digital technology which makes efficient (flexible) use of frequencies available. Frequencies are a scarce resource and governments in OECD countries have not developed a uniform approach in allocating frequencies for telephony. GSM technology and even more advanced third generation UMTS technology is helping to reinvent the phone network, while also pointing to the special aspect of network effects which are well known in telephony. The new element in mobile telephony is that access to the users is very flexible and does not need very high fixed investment (sunk costs), while a base station of a cellular network has some cost degression. If there are many subcribers in the area the phenomenon of falling average and marginal costs is much less a problem than with fixed-line networks. Hence the problem of natural monopoly is less relevant, so that there is little reason why government should interfere in price setting. Competition works; it would work perfectly, if there would not be the problem of linking new mobile networks to the existing fixed networks.

Network effects are an interesting phenomenon in telephony and the use of the internet, but also in technical standardization. The existence of network effects in the early stage of network expansion means that the marginal utility of user i is increased if there are more users (j,k,l,m) also using the network. With more users linked to the network there is an exponential increase in the number of potential links and thus the combination of bits of information.

The network phenomenon is also observed in software standardization and the use of networked x-compatible computers. This points to the importance of first-mover advantages, those which adopt the new network/software/standard early on stand to benefit strongly from the process of network expansion. First mover-advantages are even more important in a dynamic technological environment.

Building efficient networks increasingly requires one to focus on international markets and to consider foreign direct investment (FDI). However, acquiring stakes in foreign competitors and rolling out new fixed or mobile networks abroad is tightening the financing constraints of most „national" EU telecoms operators. As a consequence of building up international network structures, and due to privatization, the ratings of most telecom operators have fallen. While larger networks adequately reflect the rising market radius and the internal EU market, the cost of capital has increased. Thus the benefits of market integration come at some (marginal) costs, and with some effort it is certainly possible to quantify the costs of integration in the sense of establishing networks which cut across different countries. This political aspect must be distinguished from a pure national network expansion problem which one faces under uniform political side-constraints, ie in the US if a regional network operator is expanding into other regions of the country. From the perspective of a FDI host country the inflow of foreign investment is often desired because of technology transfer. Ie many national operators in EU countries (including those formed from regional operators (case of Italy and Denmark)) had to be restructured and governments were eager to obtain a foreign strategic investor who could undertake the required restructuring quickly; US operators were quite active in this respect in the late 1990s in Europe. Belgacom in the fixed-link telecoms segment in Belgium is another example, as is Bell South, which helped Mannesmann (acquired by Vodafone in 2000) to become the No. 1 mobile operator in Germany in the late 1990s.

Mobile penetration rates have increased by some 70% in Western Europe in 1998 and 1999, but they will slow down as high penetration rates of more than 50% have been achieved at the beginning of 2000: Finland 72%, Sweden 64%, Norway 64%, Denmark 59%, Italy 55%; the UK and Spain had 50% with France and Germany following with 40%. Italy has a low computer density among EU countries which naturally implies problems in the growth of internet use. However, among the large EU economies the Italian mobile penetration rate is the highest, which might partly offset the backwardness in computer density. If mobile internet applications would become more widespread with advanced UMTS technologies, Italy might have a relative advantage in catching-up with the UK, Germany, France and the Scandinavian countries.

The expansion of the internet has just begun here at the beginning of the 21st century. While computer networks and modern telecommunications already existed in the 1970s and 1980s, the internet is a special feature of the late 1990s. A remarkable feature is that it creates a digital world economy in which there are true world markets for many products and services. The main aspects of the internet at first glance are:

- creation of larger markets since transportation costs are falling;

- savings on transaction costs and hence a stronger tendency for outsourcing which should enhance creation of new firms;
- removing market demarcations so that there is more competition in certain fields: with the digitization of voice, music and pictures formerly separated markets are converging;
- digitization allows many product innovations and encourages novel service concepts which should stimulate entrepreneurship and economic growth.

Computer networks, modern telecommunications and internet make the economy more flexible and productive. However, to reap the productivity gains workers from many firms in declining sectors of the old economy have to move towards the expanding digital sector and this is difficult to organize. In the EU with its rather rigid labor markets it is certainly more difficult to achieve than in the US, where factor markets are more flexible than in Europe. Among EU countries the UK stands out as an economy with a relatively flexible labor market and a highly developed capital market. However, the UK is obviously not the EU's leading digital economy if one is to judge by the intensity of internet use among households or firms (demand side of the internet economy) or the growth rate of internet hosts (supply side of the internet economy). Since the use of computers in the UK is not very different from the US there must be special impediments for the non-leadership of the UK. As we will see the main problem in the UK is insufficient competition in local telephony. From the perspective of the internet, local telephony (giving customer access) is most crucial.

Fig. A12: Internet Host Density and Server Density in Selected OECD Countries

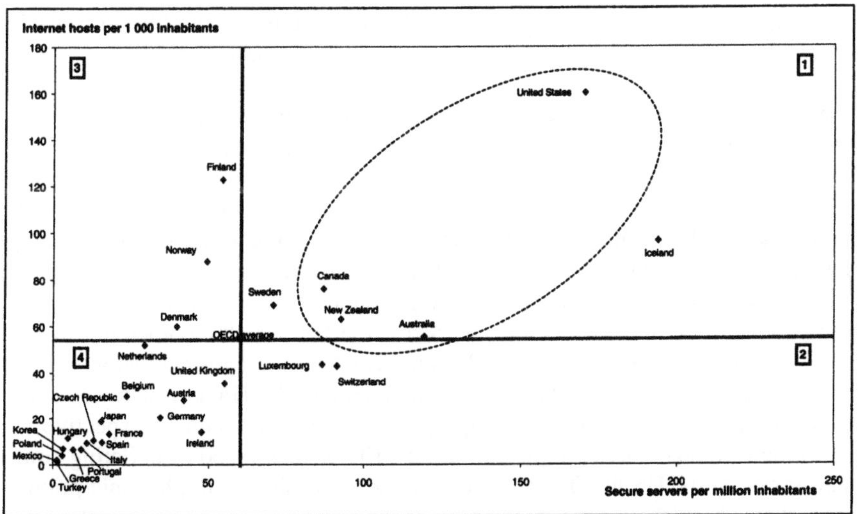

Note: Data on secure servers are from March 2000. Data on Internet hosts are from September 1999.

Source: OECD (2000), Local Access Pricing and E-Commerce, Paris, p. 32

At the beginning of the 21st century the OECD countries are indeed characterized by a digital divide on top of the „traditional digital divide", which refers to the digital haves and have-nots in society; that is those with and without access to and experience with the internet. In a recent report the OECD (2000) has pointed out the problems of a digital divide within OECD countries where the OECD study points out that there are four rich OECD countries, US, Canada, New Zealand, Australia (and the relatively poor Mexico), with unmetered local pricing in telecommunications which stimulates the expansion of internet strongly. Among OECD countries with metered local telephony, Sweden and Ireland have very low rates and therefore it is not surprising to find that both countries, plus the above mentioned wealthy four OECD countries, have host densities and server densities above the OECD average (see Fig. A12).

Local Telephony

Unmetered local telephony could become one of the crucial elements in explaining growth of electronic commerce, with respect to B2B (business to business relations) and B2C (business to customers), as well as T2S (teachers to students). The other crucial aspect will be speed, and here market structure at the local telephony level plays a role once again. Fast digital subscriber line (DSL) technologies will allow transmission of many more signals per second to customers – TV-type pictures can then be used for digital business. Pictures can be quite crucial for certain services, for marketing and customer support, but also for modern teaching. Indeed, for the whole music, multimedia and movie industry production and distribution will be much affected by the spreading of fast internet. As we will see, the former national telecomm operators in the EU countries could be formidable barriers for a broad and cheap rolling out of more bandwidth and faster transmission.

In combination with the increasing use of computers and the, internet the dynamics of the telecommunication market is crucial for economic growth. Faster and larger communication networks, declining transactions costs and new value-added services contribute to productivity increases, innovation, growth, structural change and employment. The internet and mobile telecommuncations are creating a new technological and commercial battle field in the triade and worldwide.

The leading "New Economy", combining computer use, internet dynamics, and strong competition in the telecomms market is the US, which recorded not only a record economic expansion in the 1990s but also a surprising acceleration of productivity growth in the second half of the economic cycle. High growth was combined with low inflation and full employment. In a transatlantic perspective an interesting difference between the US and the EU is the method of deflating computer prices; in the US the Bureau of Economic Analysis is using hedonic pricing approaches which show a much stronger decline of US computer prices than the EU statistics would show under identical market conditions. While different statistical procedures are observed in the EU and the US, it is not true that this explains the transatlantic growth differential observed in the late 1990s and the beginning of the 21st century. The US lead in the period 1996-2000 is 1.5-2 percentage points, on a per capita basis slightly more than 1 percentage point. Since the share of computers in total EU output is much smaller than in the US, the different statistical approaches to deflation computer prices cannot explain the difference. Applying US procedures to Germany would raise the growth figure for Germany by about 0.1 percentage points, and the order of magnitude in the EU is likely to be similar. Nevertheless one should adopt hedonic pricing functions in EU statistical analysis soon since the quality improvements in the computer industry are so dramatic.

While labor productivity in the US has certainly benefited from rising education expenditures in real terms and some capital deepening (including increasing use of computers), there has also been technological progress which was strongly influenced by information technology. In the four years ending in 1996, productivity growth increased by 0.9 percent per year, however, in the four years ending in the third quarter of 1999, it averaged 2.7 percent per year (BLINDER, 2000). While the US has seen similar spurts of productivity growth in the past (1990-92, 1983-86 and 1977-78), these accelerations followed recessions after which firms found it easy to raise productivity as a consequence of laying off workers in the recession and rising demand in the economic upswing. So the US productivity surge is, as emphasized by BLINDER (2000), not part of bouncing back from a recession. Part of the explanation for this strong increase in productivity could be internet expansion, or higher innovation dynamics, which has allowed firms to easily link with each other and firms with new customer networks. Interconnection is the key here, and network effects could be important. If the „internet hypothesis„ is correct, the main productivity effects of the strongly increasing use of computers would not be brought about by the computers alone, but partly or even dominantly by networking effects. The use of telecommunications thus plays a crucial role here.

Germany (and the EU) is lagging behind the US not only in terms of internet use and computer penetration, but it was also slow in efficiently liberalizing the telecommunication sector. Under the Kohl government Deutsche Telekom AG was partly privatized in the mid-1990s and, with Ron Sommer later taking over as the new CEO, heavily restructured and pushed toward new business fields. Germany was thus catching up with the UK where government had already allowed for competition in 1984. Germany followed the US, which had introduced long term competition in international telephony after the modified judgement of 1982 which dismemberd AT&T into regional baby bells maintaining local and regional telecommunications. The new AT&T stayed in long distance and international telecommunications and had to face Sprint and MCI as two increasingly powerful newcomers. By the end of 1999, AT&T´s market share had reduced to about 55% in long distance telephony in the US. Meanwhile, AT&T has acquired cable TV networks and mobile telecommunications so that is has become a partially vertically integrated modern telecom firm which can offer transmission, value-added services, and all kinds of internet activities. However, the market for consumer communications has become rather unprofitable in the 1990s and certainly offers weaker market growth than business services, where the demand for fast data transmission and sophisticated value-added services is increasing strongly. In summer 2000 both AT&T and WorldCom, pointing to consequences of its failed merger with Sprint, announced plans to spin off the consumer business as a separate company and BT followed with a similar announcement, although technically it would be difficult to demerge the integrated US or British network.

A major motivation behind the ideas of splitting off consumer telecommunication network operation and services is the relatively slow growth which undermines the prospects for real stock market price increases; only healthy growth in combination with rising profits promises strong stock market performance. Firms in the US and elsewhere are only too aware that internet newcomer company AOL´s taking over of the established media giant TIME WARNER could signal that fast growing newcomers, benefiting from very high price-earnings ratios, are well set to take over established slowly growing giants. Rapidly growing and fast internationalizing AOL, which, with the takeover of TIME WARNER, is attempting to become a worldwide integrated value-added company based on the global internet, is a prototype of a flexible and innovative company. AOL, together with its subsidiary Compuserve, which targets the business community, is one of the future large players in the global internet community. In Germany the growth of the internet is impaired by relatively high prices; a recent study showed that introduction of low flat rates could generate considerable growth and employment effects (WELFENS/JUNGMITTAG, 2000). There are favorable prospects for high growth of e-commerce in the global economy (MANN/ECKERT/KNIGHT, 2000).

When AOL announced in September 2000 that it would sell part of AOL Japan to NTT DoCoMo and support the expansion of the i-mode mobile internet technology developed by DoCoMo, the global market structure of telephony had decisively changed since the world´s leading internet provider had entered the network business.

As regards internet-based commercial activities one must distinguish between transactions between business and consumers (B-2-C) and business to business (B-2-B). While the consumer internet market is rapidly growing and prices for network use is falling, reflecting worldwide increase of competition in telecommunications, and innovative services are being developed, the most promising market for e-commerce concerns wholesale transactions between businesses. In the US „B-2-B“ market one expects, according to the 2000 Economic Report of the President (COUNCIL OF ECONOMIC ADVISERS, 2000a, p. 116), that sales in business-to-business will increase from $43 billion in 1998 to over $1.3 trillion by 2003 (sales is, of course, not identical with value-added).

There are several business sectors which have already cut transactions costs and broadened the international customer base by using the internet. DELL (computer manufacturing), CISCO (networking technology) and AMAZON (electronic selling of books and other items) have become well-known examples in the US. In Germany INTERSHOP, RICARDO, BROKAT, and SAP are well-known examples of internet and internet-based software companies. Even more exciting are plans of „old industries“, such as the chemical industry or the automotive industry (ie DaimlerChrysler in Germany and the US) to build electronic market places which could save the company billions in costs for procuring intermediate products and parts.

Building global electronic marketplaces will make demand functions more elastic so that price competition will become more intensive. In a vertical industry perspective suppliers stand to lose in the new digital game, the producer at the end of the pipeline (car manufacturers in the case of the automotive industry) stand to gain. Firms in the OECD area which face a more intensive price competition will naturally try to improve the position in the market by launching product innovations. Hence the internet should stimulate the innovation process.

The digitization of the economy and the whole political and social system will create a different system which is more complex, more innovative, more flexible and more competitive. Telecommunication network business will strongly benefit from the rising demand for network capacity, since conveyance of digital signals, reflecting voice, audio, or video, is important for communication and modern value-added services. The latter concerns remote services on machinery and equipment, houses and cars, but also distance learning and e-democracy in the future. Growing network capacities are based upon rising investment in telecommunications. Increasing competition in telecommunications, stimulated by privatizations and deregulations worldwide, has contributed to a fast growing rate of patent applications and a growth in innovative services. Innovative services are quite crucial for OECD countries in which the share of industrial valued-added has fallen to below 40%, and in which the majority of jobs are no longer in industry but in services. Measuring service sector growth is quite complicated for various reasons, where the use of intermediate products is a problem for productivity measurement (TRIPLETT/BOSWORTH, 2000)

Given the specific nature of services it is clear that innovative features in this business are likely to be underestimated. Obtaining a patent for electronic services is quite difficult, and in any case more difficult than in the old industry. Internet firms´ dynamics are thus often estimated by focusing on trade marks and customer growth. Customer growth in network services is, however, very different from the traditional industry.

Network effects are quite crucial here, that is that the marginal value-added for customers is increasing if the number of links to the network is rising. More links imply more opportunities for interesting and profitable combination of knowledge, experience and capital. The network benefits in mobile communications and mobile internet business will be considerable since network expansion costs in a world of mobile internet are relatively low compared to traditional fixed telecomms networks. This will create extra profits for internet providers and for telecommunication firms over whose network the internet business is going. Large telecommunication network firms which also offer value-added services, including internet and mobile telefony, are well positioned to reap profits from network effects in the early internet age.

3. Boom in Telecommunications in the OECD

3.1 General Trends in Leading Industrial Countries

With modern digital transmission technology plain old telephony is over and a new era of communication has begun. In regards to the organization of firms, the combination of powerful computers, mobile telephony, internet expansion, and the technological convergence of voice, audio, and video means a rapidly changing industrial landscape in OECD countries. They all benefit from the strong competition in telecommunications which has not only been fostered by technological dynamics, but also by market deregulation and privatization in Europe and worldwide in the late 1990s (WELFENS/GRAACK, 1996; WELFENS/YARROW, 1997, WELFENS/GRAACK/GRINBERG/YARROW, 1999).

Telecommunications is a booming business worldwide for several reasons:

- It is characterized by high technological dynamics, which means that technological leadership, flexible organization, and timely adjustment are crucial.
- The intensity of competition is growing as successful newcomers from mobile telephony are entering the fixed network business. A prime example in Europe is Vodafone whose German fixed network subsidiary Arcor is expanding in Germany´s fixed link market; in the UK, where Vodafone had offered jointly with Energis PLC and Cable & Wireless a fixed link service only to corporate customers, Vodafone decided in October 2000 to offer jointly with Atlantic Telecom (a US company which had bought First Telecom, an aggressive British fixed network operator, in 2000). New wireless local loop technologies will be employed in the UK, Germany and other countries. In Germany, unbundled access also offers the option to rent local access; in the UK the dominant BT will have to offer competitors similar access first in 2001. However, there is little doubt that the Commission will adopt a directive which will require all EU countries to offer unbundled local access.
- Computer business, the internet and telecommunications are expected to be a very profitable business in OECD countries. High rates of return will, however, go along with a high variability of profit rates. Rapid growth of mobile telecommunications worldwide stimulates in particular the demand for chips (used in mobile phones), while the rising growth of the internet raises demand for computer servers and chips. If many users from European city X want access on a US website with streaming video at the same time, there will be long delays for European users. It would make sense for internet service providers to download the streaming content (say a new movie) to a server close to the main European users; the file would be copied across the Atlantic within a second, and waiting time for European us-

ers would be cut dramatically. This shows that the internet is a worldwide web but proximity of servers and customers can play a role for premium services.

- In a period of globalization, namely growing trade in goods and services, as well as capital flows, the need for optimal information sourcing, processing and storage has increased dramatically. Moreover, with innovation cycles shortening, optimal innovation and diffusion are important, and here modern telecommunications and internet play a decisive role.
- It is rapidly internationalizing as national monopolies in Europe have been removed and as the optimum firm size is growing in an enlarged market. Moreover, leading multinational companies (MNCs) in OECD countries need one-stop shopping for their company´s telecommunications services, and large MNCs with large budgets know markets from a comparative perspective and are mostly organized in the Telecommunications User Group.
- It is going mobile increasingly where digital cellular subscriber density in Scandinavian countries is about to match that of fixed network density. Mobile internet applications will become popular with GPRS and UMTS (3G) technologies.
- It should benefit from the competition stimulus from technological convergence which reflects the fact that digitization (substitution of analogue by digital transmission technology) has removed traditional market demarcations between TV, telecommunications (voice telephony) and data transmission.
- It is a key for the internet expansion, and in particular here of so-called network effects, which means that early users of the new net will enjoy larger benefits than new users; potentially interesting communication partners are linked up to the network. Thus the demand for modern communication services is expanding, not only because new users generate new revenue, but also because the willingness of existing users is increased as the network is expanding.

There are new value-added services which are gaining importance where internet and mobile telephony are two key areas. As profit margins in the traditional fixed network operation come under pressure, dominant fixed network operators naturally start to expend new services. Internet and mobile telephony play a key role. Often customers can buy only bundled services, which impairs competition to the extent that considerable cross-subsidization makes market entry of newcomers into selective markets almost impossible.

Problems of Sustained Competition in Digital Services

Liberalization of fixed network telecommunications worked quickly in Europe as most economists had expected in a dynamic technology field. Prices came down, services improved and gradual or full privatization brought extra revenues for ministers of finance in EU countries who were all eager to meet the Maastricht deficit criteria.

Local telephony is not as competitive as national and international long distance calls which faced more newcomers and easier market entry. The local loop is difficult to conquer as laying a parallel network to get customer access is rather costly, but the technical opportunities to unbundle local access, that is split the line into a voice telephony access and an internet access line could stimulate internet expansion where the alternative for customer access is via satellite or cable TV. Wireless local loop is another option applied in some European countries.

The European Commission proposed to unbundle the local loop in 2000, but changing business in the local loop takes time as can already be seen in the US. While the Commission Directive will force all EU countries to unbundle the local loop and to allow newcomers to rent local access lines from the incumbant operator, this is only part of the necessary measures to create sustained competition. The main problem is that ex-monopoly operators have increasingly started to bundle regulated and unregulated products which implies the transfer of a monopoly position in one market, typically local access and local telephony, into other markets. The FCC (KENDE; 2000) and the US government in the AT&T antitrust case (LAFFONT/TIROLE; 2000) have always emphasized the importance of avoiding the transfer of market power one market to the other. As regards the EU the post-1998 development in digital services is problematic insofar as the incumbant operator has a tendency to bundle regulated and unregulated products. While normal bundling in markets is often accepted by competition authorities the problem here is that a dominant operator is defining what normal bundling should look like. Competitors and newcomers will have to offer the same or similar bundles as the former monopoly operator, not least because private household as well as firms like one-stop-shopping because of convenience and price advantages. While in EU countries the incumbant operator will typically argue that competition in telecommunications (only voice telecommunications and interconnection typically is regulated, and sometimes the rental price of local lines) has increased over time (and falling prices partly support this view), the reality is that there is a transitory increase in competition which would fade out as soon as regulation is abolished. The initial fall of prices will clearly reflect the one-off effects of opening up, including removing X-inefficiencies from the incumbant operator. The medium term fall of telecommunication prices largely reflects the effect of high technological progress in the field of telecommunications. However, only a few years after opening-up of markets there is a clear tendency in EU countries for ex-monopoly operators to start to offer bundled products which combine regulated services and unregulated services.

3.1.1 Privatization

Privatization of telecommunications in an environment of market opening is a difficult task. Once partly or fully privatized, the former telecommunications monopoly operator will be eager to increase profits, despite the fact that market opening will naturally bring falling prices and falling market shares for the incumbant operator. A joint stock company will, however, use every opportunity to increase profits and the share price of the company – this means strong incentives for attempting to erect new barriers to entry.

Privatization has the advantage of making the company really transparent, since there are some minimum requirements to be fulfilled in order to obtain a stock market listing. Privatization raises some complex transitory problems since the management of the privatized firm will probably want to cut the labor force, although many, if not most of the workers and employees are civil servants and thus cannot be easily dismissed. Management´s key challenge it to raise productivity, to come up with new innovative services and to maintain the top rating the company had when it was still a government-owned company. Moreover, the management has to explore the links between the changing market size, which is increasing, and the optimum firm size.

There are many arguments for privatizing state-owned network operators, which normally means launching an initial public offering. The initial offer price should not be too low because the Minister of Finance will face complaints that the „crown jewels" were given away. The initial price (offer range) must leave some room for capital gains because this will help to establish a winner reputation for the telecom's privatization. Given the enormous relative size of the telecom operator (often No. 1 or 2 in terms of stock capitalization after the IPO), government will normally organize privatizations in various stages. Taking this into account also means that one should aim at conditions leading to modest (if not considerable) capital gains in the first and second privatization round. The size of the telecom operator will normally make it one of the leading stocks in the leading stock index which also implies some prospect for capital gains. What could, however, be a serious problem is when the privatized telecom operator faces medium term limitations for full privatization and/or weakness in international telecommunications. The prime example here is the problems from the Swedish Telia´s flotation in 2000. The government had a mandate to sell up to 49% of the government stake in the company, which is considered to be one of Europe´s leading companies in mobile telephony. The government sold 30% of the shares to the public in June 2000, but in August the share price stood at 66.50 SKr, which is about 1/5 less than the launch price of 85 SKr. While the stock traded at almost 95 on the first day it then fell back as a consequence of several influences: International investors sold shares (their share fell from 36% to 28% within 10 weeks), and some domestic investors became skeptical when they became aware of both the restructuring needed at Telia and the increasingly difficult situation in EU mobile telephony, as evidenced by the high $46 bill. revenue from Germany´s 3G mobile auction. Neighboring Sonera from Finland had put itself up for sale in mid-2000 as it considered itself to be too small to survive on its own in the increasingly international telecommunications market. Another problem is that Telia is a latecomer in the EU privatization process, so that one must assume that other companies have made faster progress in restructuring.

Investors in Spanish Telefonica and German Deutsche Telekom have also faced considerable volatility in their stock price. Based on the offer price, Deutsche Telekom doubled its share price twice, but fell to ½ of its peak in 2000. The pressure of the stock market will certainly drive privatized telecom operators to restructure and to become more innovative, and the development of patent applications immediately prior (in anticipation of privatization) and in the years after IPO confirms this hypothesis for most telecoms operators in Europe. However, one must emphasize that stock market volatility of US telecoms shares was also relatively high in the 1990s.

3.1.2 Some Regulatory Problems after Privatization and Opening-up

As long as government owns a monopoly telecomm operator, it will regulate tariffs in a way which meets certain political and economic criteria. However, once the market has been opened and partial or full privatization is underway, there is a need to regulate the telecommunications market in a new way. There is need for regulating

- access and access fees
- interconnection and interconnection fees
- universal service obligations

If one has opened up the telecommunications market the typical situation will look as follows: The ex-monopoly operator can rely on a majority of traditional users with a preference for the incumbent operator. Certain strata within the existing demand side (probably mostly young, urban, well-informed users) will defect. The introduction of competition can then be displayed as follows (see graph). In panel a) there is an initial demand curve DD_o and a marginal cost curve of the monopoly operator k'^*_o, so that profit maximization (assuming that there is no regulation) will result in the point C and the monopoly price p_o^*. In panel b) we show the marginal cost curve of the newcomer group as k'_o and the defecting market demand equivalent to DD'_1, so that the incumbent operator in panel a) will face only a residual demand of DD_1. Without regulation the former monopolist will impose the new Cournot price p_1^*, which is considerably higher than the price charged by newcomers, namely p_0.

If, however the incumbent operator is supplying network capacity and transmission services at monopolistic prices to newcomers their marginal cost curve will be shifted upwards by a factor z so that the newcomers' price now is p_1. As the incumbent operator is facing gradually increasing competition the marginal cost curve in panel a) will shift downwards, say to k'^*_1 so that the premium charged by the incumbent operator has reduced to the difference p_2^* minus p_1. Oligopolistic modeling could be interesting here - but we will not explore this field.

Under full competition there would be, of course, a uniform price. If there is a regulatory authority which imposes price regulation one may assume that the regulator will set a regulated price for the incumbent operator below the monopoly price. There will be an excess demand in market a) which is represented by consumers and firms which will ultimately switch towards the newcomer market segment so that the demand curve in panel b) is shifting outwards over time (not shown in the graph).

Fig. A13: Residual Monopoly and Opening up of Market

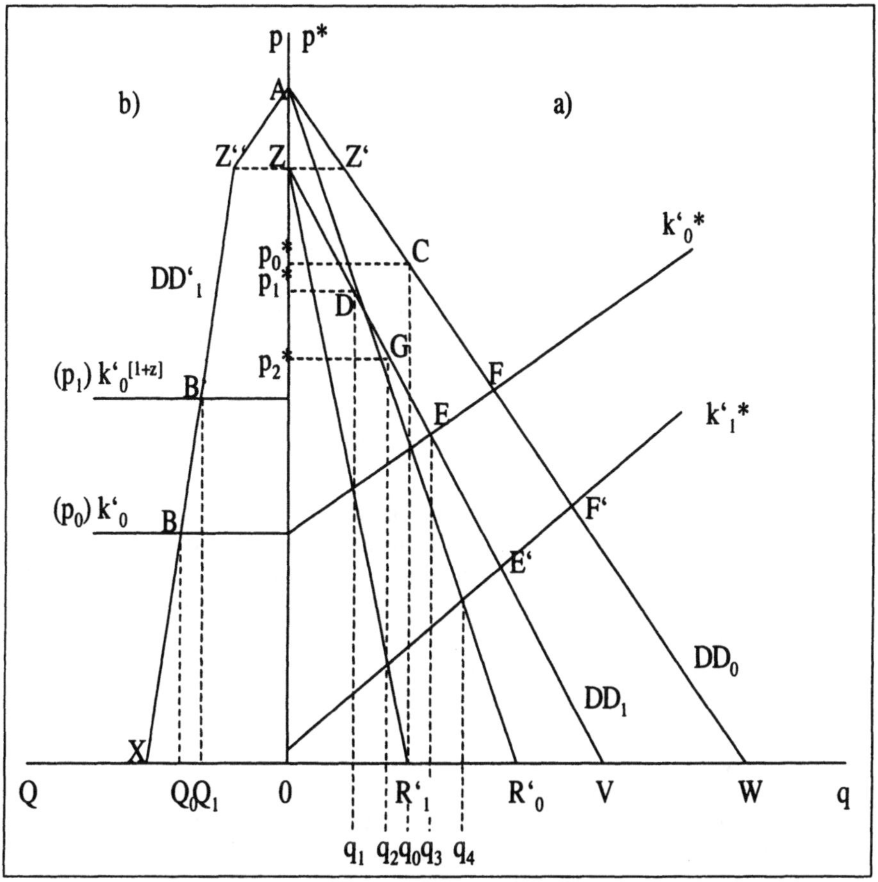

3.1.3 Mobile Telecommunications: Growth and Auctioning of Licences

Mobile telecommunications is a new and dynamic growth field in Europe. In 1999 Finland and Norway had penetration rates close to 60%, Italy had 52.5%, Japan 44.9%, the UK 40.3%, France 34.9%, the US 31.5%, Germany and Canada 28.6% and 22.7%, respectively. Between 1995 and 1999 the share of wireless subscribers as share of populations in G7 countries and Scandinavia has more than tripled as is shown in the following table.

Tab. A11: Mobile Penetration Rates in G7 Countries and Scandinavia (%)

	1995	1998	1999
Finland	19,9	58,0	65,0
Norway	22,6	48,6	61,8
Sweden	22,8	46,5	57,6
Italy	6,9	35,8	52,5
Japan	8,2	37,7	44,9
UK	9,8	25,6	40,3
France	2,5	19,1	34,9
U.S.*	11,8	25,5	31,5
Germany	4,6	16,9	28,6
Canada	8,5	17,8	22,7

* US figure for October 2000 exceeded 35%

Source: OECD Telecommunications Database and US Council of Economic Advisers

In a White paper the COUNCIL OF ECONOMIC ADVISORS (2000b) notes several key points, including:

- Telecommuncations and the Internet are among the most important sectors of the New Economy. Telecommunications represented 3% of GDP in 1998, having grown at a 7 percent annual rate over the previous 10 years. Wireless carriers employ over 150,000 people in the United States and generate $44 billlion in annual revenue. At the end of 1999, the United States had 86 million subscribers, today that number exceeds 100 million. By the end of year 2000 there will be 600 million wireless subscribers worldwide...Consumers purchased $5.5 billion in goods and services over the Internet during the second quarter of 2000 alone. Sales over the Internet between businesses are estimated to hit $251 billion in 2000, up from only $43 billion in 1998. The most successful Internet startup companies have created hundreds of billions of dollars in market value.
- Third generation wireless technology combines two powerful innovations: wireles communicatins and the Internet...
- This new technology promises substantial benefits to consumers, producers, and the economy as a whole. The annual consumer benefit from today´s wireless telephone services is estimated at $53-$111 billion. The consumer benefits from 3G services will likely be of this order of magnitude.
- To provide 3G applications most efficiently, adequate spectrum must be made available for commercial use. In telecommunications, the most important scarce resource is spectrum..."

The COUNCIL OF ECONOMIC ADVISERS (2000b, p.8) concludes on the basis of existing 2G wireless industry in the US that under the given spectrum allocation (valid in 1999) the average US wireless operator earned $0.238 bill. of revenue per MHz; therefore an additional 150 MHz of spectrum could generate an additional $35.7 bill. of service revenue per year. The price of $22 at off-peak times for wireline Internet access in 2000 (compare to the OECD average of $35) is relatively low so that interest in wireless internet use might be rather limited in the US. Finnland, Norway and Sweden had a wireless subscriber share of 3-4%, Japan was leading with 7.9%. The Japanese i-mode (mobile) internet access thus is raising the overall level of internet penetration in Japan; wireline internet access amounted only to 20.8% in mid-2000. Sweden and the US were clear leaders with 50%, Norway reached 49.4%, Finnland 42.6%, Italy and Germany nearly 20%. Canada reached 41.8%.

Tab. A12: Wireless Subscribers, Internet Access, and Wireless Internet Access as a Fraction of the Population in G7 Countries and Scandinavia

	Mobile phones, year-end1999	Internet access, mid-year 2000	Wireless Internet subscribers, 2000 estimates
Finland	65,0	42,6	3,7
Norway	61,8	49,4	3,4
Sweden	57,6	50,3	3,5
Italy	52,5	19,6	2,1
Japan	44,9	20,8	7,9
UK	40,3	32,7	1,4
France	34,9	11,0	1,5
U.S.	31,5	49,8	1,3
Germany	28,6	18,0	1,2
Canada	22,7	41,8	N/A

Source: COUNCIL OF ECONOMIC ADVISERS (2000b), The Economic Impact of Third-Generation Wireless Technology, White Paper, Washington DC, p.16.

The growing economic significance of the mobile phone market in the US can be recognized from the enormous increase in cumulated capital investment of US wireless carriers which reached $0.9 bill. At the end 1985, but $71.3 bill. by the end of 1999 (COUNCIL OF ECONOMIC ADVISERS, 2000b, p. 13). EU carriers are likely to roll out G3 mobile technology faster than the US carriers which face problems on two fronts. Almost half of all US wireless subscribers in 2000 were still using analogue transmission technology, while the percentage in the EU is above 90%. In the US TV companies and the military hold much of the optimal spectrum needed for rapid and cheap building-up of 3G mobile networks. By contrast, in most EU countries 3G auctions were held in 1999/2000 so that rapid building of 3G networks can be expected in the EU. This modern network is a natural exention of the existing digital GSM network in the EU.

UMTS Licences and Market Dynamics in Germany

Governments in most EU countries decided to hold an auction for awarding 3G(UMTS) mobile licenses. The Netherlands raised 2.7 bill. Euro, Italy 11.8, the UK an impressive 38.5, and Germany an unprecedented 50.8 bill.; the results in Italy (and in Austria) were rather disappointing. Since frequencies are a scarce resource it is obviously adequate to hold an auction; the alternative would be a kind beauty contest (applied in Spain and partly relevant in France) which, however, smacks of political favoratism and also has lack of transparency. The price companies are willing to pay critically depends on discounted future (after-tax) profits; from this perspective expansion of UMTS networks in Germany and the UK seem to be particularly profitable.

In the following paragraphs we will take a closer look at the successful German auction; before we do so we will take a brief look at the market situation in Germany. Within less than three years competition in Germany has intensified strongly as is shown by rapidly falling prices (almost 80% in some segments) and rising patent application figures at Deutsche Telekom AG; the incumbant operator thus has reacted to the new regulatory environment and the gradual privatization of the dominant telecom operator. The success story of German deregulation and privatization has been relatively impressive in long distance and international telecommunications mainly for six reasons:

- There were aggressive newcomers, Mannesmann (D2) for instance, which used its sucess in mobile telephony to move into the fixed network business as well. Mannesmann benefitted from its alliance with Bell South which took a minority stake in D2. It is, however, true that the most interesting option for launching a new fixed network, namely using the cable TV network which covered about 2/3 of the population was not used since the minister of Finance (then Theo Waigel) stubbornly refused to hive off cable TV from the Deutsche Telekom.
- There were powerful newcomers from the energy sectors when three major private energy producers entered the telecomms market by opening up internal telecommunications networks, investing in new capacity and using experienced billing procedures for the new field of telecommunications. While they did not survive in the telecom business (not really surprising as they came from a sheltered market with modest technological dynamics, which is just the opposite of the traits of a deregulated telcomms market), they created markets and customer groups for domestic and foreign newcomers which took over the telecomm's activities from the energy firms. This is in market contrast to France where France Telecom faced no competition from state-owned monopolist Electricité de France. In this perspective France was poorly deregulated, much worse than Spain and Italy where energy suppliers moved into the telecommunications business.
- The addition of 1/5 of the former population as a result of German reunification in 1990 (with a phantastic excess supply) resulted in an enlarged German market with a large strata of population with weak customer loyality to the dominant operator. Probably the biggest counter-example is the UK where British traditionalism seems to have impaired consumers´ willingness to explore new opportunities for switching to newcomers (there has been full competition in the UK since 1991).
- Aggressive retailers have gained market shares of several percents and generated revenue that allowed them to build core capacity which could be combined with leased lines.

- The regulator has been rather consistent in its support for newcomers and in helping to avoid a general dominance of the incumbant operator.
- Germany´s mass media (mainly private TV stations and newspapers) competed vividly not only for advertising for all the new companies and services, but they engaged in semi-professional regular price comparisons of telecommunication services as well.

Prices have fallen sharply in long distance and international telecommunications, while competition in the local loop is unfolding more gradually as options for bypassing the fixed network are expensive unless a modern cable TV network has been laid. Here Germany has a major problem since it is the only country in the European Community, except for Portugal, where the dominant fixed network operator also owns the cable TV network. Deutsche Telekom AG is facing a critical debate from the Monopolkommission (Council of Competition Experts) and academia as to why such a double monopoly should continue to exist. The TV set--top box used by Deutsche Telekom, which is not an open platform for pay TV, is also under scrutiny.

A major problem of competition in Germany concerns local telephony which is particularly critical for the internet. Here the European Commission proposed new directives in mid-2000 (still to be adopted by the European Parliament) which would unbundle the local loop so that newcomers in Germany could not only rent local access lines, but also unbundle local access in the sense of renting a certain transmission capacity. This clearly would stimulate the expansion of the internet, which has been hampered seriously by very high prices in Germany in 1998/1999. It was only in spring 2000 that Deutsche Telekom announced the introduction of a monthly flat rate of DM 79,-- ($39) which was immediately countered by AOL with DM 78. In late 2001 AOL was still critical of Deutsche Telekom since the dominant operator refused to offer a cheap wholesale flat rate for internet providers as it exists in the UK and the Netherlands since 2000. In the UK, the national telecomms regulator OFTEL had imposed it and thereby cut the price of intermediate input for internet service providers considerably.

The German regulator is responsible only for telecommunications, but not directly for regulating the internet, although whatever is regulated in the local telephony market clearly will affect the internet market, where providers need the direct access to each customer. The regulator holds the view that a wholesale flat rate effectively means that rare internet users will subsidize a small minority of internet freaks. This opaque perception ignores the fact that a flat rate on all layers of the network will not only generate positive external economic effects but also bring down metered internet prices.

Deutsche Telekom has partly yielded to public and political pressure and started to sell minority stakes to strategic outside investors which are mainly from the US and the UK. Besides price competition there is an increasing Schumpeterian competition, namely a wave of innovations including novel services related to the internet. The number of patent applications of Deutsche Telekom AG has steeply increased in the second half of the 1990s, which partly reflects a new management style geared towards technological progress and international competition. Sales proceeds from the cable TV business and cash from new public offerings in 2000, including a separate initial public offering for the internet subsidiary T-Online, have allowed Deutsche Telekom to reduce the debt-sales ratio and to envisage major international mergers & acquisitions. This is certainly needed given the rapid globalization of the telecomm market after the collapse of cooperation (associated with symbolic minority cross-holdings) with France Telecom in 1999, and the failure to take over Telekom Italia in 1999.

G3 Mobile Telecommunication Auction in Germany with Record Revenues

In Germany, the auctioning process for third-generation (3G) mobile telecommunication licenses began on July 31, 2000, when seven bidders entered the race. The auction ended after 173 rounds of bidding on August 17 with six bidders left (DEBITEL had pulled out) and record revenues of $46.5 billion (close to DM 100 billion) in the EU. The British auction held a few months earlier had brought in $33 billion. On a per capita basis, the figure for Germany, namely 615 Euro, was close to the British result of 630 Euro. The difference is insignificant if one concedes that the Euro equilibrium exchange rate has a „natural floating margin" of +/- 10 percent.

The government had offered 4-6 licenses valid as of 2002 for Germany, with each license valid for twenty years. Bidders need either two or three blocks to obtain a license. However, it turned out that each of the six bidders bid for only two blocks, where Deutsche Telekom (actually its subsidiary T-Mobil) offered the highest price, that is, 16.58 billion DM. Viag Intercom (owned by BT and Telenor) got the license for 16.52 billion, Mannesmann Mobilfunk (Vodafone AirTouch) for 16.47 billion DM, Group G3 (Sonera/Telefonica) for 16.45 billion, E-Plus (Hutchison/KPN) for 16.42 billion, and Mobilcom (backed by France Telecom) for 16.37 billion. The only newcomer among the six was the Finnish-Spanish consortium Sonera/Telefonica. While incumbent operators face additional investment costs for rolling out the technology of some 10-15 billion DM, Sonera/Telefonia could have considerably higher costs, not least because in a market with about 50 percent penetration rate by 2002 it will have high marketing costs.

Hutchison Whampoa left the E-Plus backing group in less than half a day because of disappointment over the high licensing costs, but KPN's partners NTT DoCoMo of Japan and Bell South declared that they would increase their respective share correspondingly. The six bidders entered another German auction on August 18, 2000 for licenses for the transfer of data. This auction shows that the auctioning of scarce resources is becoming a normal part of telecommunications, which used to be a static government owned monopoly in most EU countries in the 1990s. Governments should indeed consider the instrument of auctions more often—the late WALRAS would be quite happy to witness frequency auctions in Europe, fetching even higher revenues than in the U.S. A major problem in the U.S. is that digital TV companies and the military have blocked some of the most interesting parts of the frequency spectrum. However, the recent results from Europe will drive the FCC in the U.S. to reconsider the problem of 3G frequency auctions, a problem that it has been slow to deal with.

After news broke about the auction result, stock prices fell by 1 to 6 percent, reinforcing earlier losses. From a theoretical point of view, the auctioning of licenses in an efficient manner amounts to taxing away the monopolistic profits that the companies would otherwise make. For the German Minister of Finance, the 100 billion DM in license fees are a much welcome windfall revenue which will probably be spent on reducing government debt, which stood at 60 percent of GDP in 1999. Fully using the windfall profit to reduce government debt will lead to government savings on future interest payments of roughly five billion DM. The federal government is willing to spend these savings for raising expenditures on R&D and on investment in the railway system. Some special government packages for eastern Germany's high unemployment regions also seem likely. Given Germany's decade-long under-spending on education, research and development and other deficiencies, taking all of the windfall revenues from the auction to repay government debt is non-optimal—to avoid the word unprofessional. However, Germany has a tradition of choosing political heavyweights with no economics or business background for the helm of the Ministry of Finance; this is in stark contrast to the tradition in the U.S., the UK, the Netherlands, Switzerland, and many other OECD countries. It is thus no wonder that Germany's growth and employment record relative to the U.S. and other countries has been so poor in the 1990s.

With five to six companies ready for further large investments over the period 2000-2005, the net investment-GDP ratio will go up by 1-2 percentage points in Germany. This should contribute to higher growth. The share of mobile telephony revenues (including new mobile Internet applications) will climb from about 25 percent in 1999 to about 40% percent in 2005. By the end of 2001, mobile 3G auctions had been completed in Austria, Belgium, Denmark, Switzerland, and the UK, the Netherlands, Italy and Germany, while France realized a mix of „beauty contest" and a high license fee. Companies can recover high license fees only by generating high revenues in G3 mobile telecommunication. This will mean that they will compete for a quick and more or less European roll-out of the new networks, and they will launch new services, say in banking (online accounts, new credit cards etc.). Users in the business community and well-off households can be expected to quickly pick up the new technology, which is good news for the telecommunications equipment industry. It will also stimulate expansion of mobile Internet activities, as G3 technology allows high rates of data transmission—even TV signals can be conveyed without problems. There is a wide array of applications in both the business sector and in households. In the medium term, Europe will lead the U.S. in mobile telecommunications. However, the EU faces the problem of market fragmentation through national political borders that will make roaming and telecommunications in the EU expensive. Only Vodafone has announced in early January 2000 that it will offer a pan-EU service at a single price.

Taking into account German figures the share of voice telephony will fall below 50% in the medium term in Europe: Based on 88 billion DM in telecoms services revenues in 1998 (REGTP, 1999, p.4) the structural breakdown was 52% (46 bill. DM) for voice telephony, 22% for mobile telephony (19 bill. DM) and 5% for Cable TV (4.5 bill. DM); cable TV business is expected to grow fast since

DT has started to sell regional cable TV companies in 2000/2001 (note that some 2/3 of all households could be connected to the cable TV network). In 2001 telecom services will exceed DM 100 bill in Germany.

The UMTS auctioning process is adequate from an economic perspective, since spectrum is a scarce resource. Allocating licenses on the basis of a „beauty contest" would leave too much discretion on the side of politicians and would also hardly make sure that the most dynamic companies would get the valuable licences, which is the key for a potentially fast growing mobile phone and mobile internet market. The main reason is that UMTS will allow very fast transmission of signals, so that even video could become a standard feature of future mobile services.

Fixed network operators which are unsuccesful in securing a UMTS network in Europe's largest economy will face serious problems in surviving the strong competition in the EU. A second-hand opportunity could, however, emerge once one of the financially weaker winners faces problems in financing the massive cost of the UMTS network.

With first auctions in the UK, the Netherlands and Germany in 2000, the rollout of the new technology is expected to be complete in the EU in 2005. The price for an UMTS license in Germany can be expected to be rather high, since the auctioning process is similar to that in the UK, where the number of licenses, namely 5, exceeded the number of incumant operators (4) so that newcomers were motivated to bid strongly for a license.

The Dutch auction was poorly managed by the government, which had offered five licenses in a setting with five incumbant mobile operators, which gave weak incentives for newcomers to bid their way into the market. Instead they teamed up with incumbant operators, something which the government´s auction rules should not have allowed. PAUL KLEMPERER (FT, 2000, July, 2000) has clearly identified this weakness and argued that adequate rules should emphasize: (i) attracting market entry, (ii) avoiding collusions, (iii) getting an adequate market structure. He argues that the Dutch model failed on (i) and (ii), while the German procedure is not meeting (iii), as the German government is facing the risk of ending with an overly concentrated market. The German regulator´s auction is for 12 blocks from which bidders can create „licenses" of either two or three blocks, which seems an interesting arrangement as the market can decide how many licenses will be created (a number between four and six), but the German rules clearly motivate players to aim at four licenses and a rather oligopolistic situation with very high profits at the expense of users. High UMTS prices will further undermine the short term profitability of DT, which already faced falling stock market prices after it indicated its firm willingness to take over VoiceStream for a high price.

Deutsche Telekom faces the problem that the price tag on third-generation mobile telephone licenses in Germany is high, but there is a clear need to invest here for a company eager to be a leader in mobile internet application. Some of Deutsche Telekom investments in Eastern Europe could generate high profits in the future. Most important seems to be mobile telecommunications and the (fixed and mobile) internet business. While user fees are likely to fall, the prospects of rising revenues from advertisement and commission fees allows one to expect that

Europe´s largest telecommunications company should earn high profits in the future. Deutsche Telekom is also likely to move aggressively into the growing international telecommunication business once it has lost its dominant market position in this field, which would free it from price regulation of the national regulator in this dynamic niche.

As regards the expansion of internet use, Germany is not among the leading three European countries, which are Sweden, Finnland, and the UK in early 2000, but Germany is gradually catching-up; user density in the business community and among private households is increasing, as is the volume of sales which, however, hardly exceeds 0.3% of overall revenues (compared to roughly 1% in the US).

Cutting internet prices by 50% could generate about 400 000 new jobs in the medium term in Germany, as reported by the EIIW study by WELFENS/JUNGMITTAG (2000). The Ministry of Economics argued in its 2000 annual report that as many as 650,000 new jobs could even be generated by a strong expansion of the internet. However, a major problem in Germany is that complementary sectors to the internet are highly regulated and therefore inflexible, expensive and slow. This partly holds for the transportation sector, and for postal services as well.

Selected Market Dynamics

The large US market has been characterized by the growth of newcomers and an increasing role of international investors. NTT of Japan wants to aquire Verio Inc., which is a dynamic US internet company with growing business in web-hosting. This has raised concerns in the US about foreign government ownership and about complication of US wiretap efforts, as the Japanese government is a majority owner of NTT.

NTT DoCoMo, as the world´s largest mobile telecommunications company, has announced plans in 2000 to take a minority investment in a joint venture with SBC Communications and BellSouth. The strategy is to combine DoCoMo´s technological leadership in the so-called i-mode of mobile telephony with the market access of two leading regional operators in the US. Within 1 ½ years DoCoMo was able to get 11 million customers in Japan for the i-mode services, which includes mobile internet services. In 2002 the US will have about 100 million mobile telecomms users and about a similar number of internet users. However, few of the latter will have mobile applications. The leading US mobile operator Verizon has not more than 2 million mobile users with mobile-internet phones. The US market is more difficult than Japan, where a higher share of people lives in urban areas, and the US market also suffers from the four incompatible competing standards in mobile telephony; worse, almost 50% of the market was based on analogue systems in 2000. With about 100 million mobile users in 2001 it will be about 50% of the European figure (250 million), but by 2005 the US figure is likely to have doubled. An akward problem for market expansion in the US is the convention that the receiving party must also pay, while in Japan and the EU the only caller must pay. US owners of cellular phones thus have an incentive to temporarily switch off the mobile phone. From a European perspective this strange US convention is an advantage for European mobile operators whose home market is thus likely to be more dynamic – and it doesn't seem very likely that the US will witness a change of this convention.

While Deutsche Telekom snapped up smaller telecom operators in eastern Europe and some developing countries, the dominant German operator was slow in its global internationalization strategy, which should include a US subsidiary. The Global One Alliance formed with Sprint and France Telekom faltered when WorldCom took over Sprint, a takeover which failed in 2000 because the FCC did not give its green light to what it considered as a competition-restricting manoeuvre. France Telecom later acquired Global One.

With Deutsche Telekom AG (DT) acquiring VoiceStream, a major mobile GSM operator in the US, for $50.7 bill., the German giant has favorable prospects for external growth and internationalization. DT argues that it can now reach a potential 375 million customers in Europe and the US.

The price tag on Voicestream is high since it implies an average of $21 000 per subscriber, which is about three times as much as France Telecom paid for Orange in the UK (Germany´s DT has a foothold in the British mobile market via its acquisition of One-2-One, here DT paid nearly $5000 per subscriber). Voice Stream has licences in 23 of the top 25 US markets, but the network is still not really dense, ie there is a strong need to invest in California. Here DT´s investment funds will allow Voicestream to quickly expand its network. At the same time DT will be eager to exploit potential transatlantic network effects, which could be based on the fact that GSM is also Europe´s and Asia´s most common mobile technological platform. DT and Voicestream will most likely heavily target international business users who will be able to use the mobile phone in both the US and Europe, as well as most of Asia. The ongoing integration of internet use and mobile telephony could generate considerable additional revenue for any global mobile operator. Here Vodafone, after taking over D2 (Mannesmann) in Germany in early 2000, is clearly number one. The largest mobile internet operator in a single national market is NTT´s DoCoMo in Japan, which has been approached by AOL for cooperation in the Japanese and other Asian markets. DT is eager to expand in mobile telephony and the internet which, will fully merge with the coming UMTS standard.

VoiceStream´s price tag is high given the fact that the company had 2.2 million subscribers in the US and about $0.5 bill. in revenue in 1999. However, the potential subscriber base is, of course, much larger. With a focus on this large potential subscriber base, DT´s acquisition might make sense in the long term. DT and VoiceStream will face stiff competition in the US, where the market leader is Verizon (using CDMA technology), which is jointly owned by Bell Atlantic and Vodafone. While Verizon had slightly more than 22 million subscribers in early 2000, AT&T, using TDMA technology, is No. 2 in the US mobile phone market with about 13 million subscribers. Hutchison Whampoa of Hongkong, which owns 22% of VoiceStream, will own 4.6% of DT shares after the takeover of VoiceStream by DT, which is paying mainly in form of an equity swap. Owners of DT have suffered a large reduction of the DT stock price since the transaction was disclosed. The conclusion to be drawn is that market entry in the US is extremely expensive, and that the tendency of EU telecommunications operators to finance takeovers via the exchange of stocks is bound to lead to sometimes interesting, sometimes conflict-prone changes in ownership.

3.2 Financial Markets and the Internet

Financial markets play a particular role in the internet age. There are three major links between financial markets and the internet:

- Financial markets transactions are easy to make via the internet so that more people can enjoy the economic benefits and the thrill of e-deals in the stock market or other financial markets. Reduced transaction costs in financial markets make it attractive for millions of internet users to conduct financial market operations directly.
- The internet has become mobile due to WAP, GPRS and UMTS technologies so that the massive growth of mobile telephone users in the OECD countries and the NICs represent a partly new group of internet-based investors; for direct brokers it is indeed attractive to team up with major mobile phone operators – the latter expecting to attract high income strata and the former new clients for direct brokerage.
- The financing of internet-based firms is an interesting challenge for stock markets and banks. With a wave of .com companies being created in the early 21st century there are new opportunities for stock markets and banks to expand. At the same time there are considerable new risks, and the anglo-saxon stock-dominated system of financial markets might be more adequate for coping with the new challenges than the systems dominated by universal banks in continental Europe.

Let us pick up the last point: The internet age is one in which barriers to market entry have reduced in the sense that the minimum investment to start a web-based firm is rather low: All one needs is a few computers, software, a telephone line and skilled personnel. However, the problems encountered in financing an internet firm are more complex than in traditional industries in which real estate and tangible assets (ie in a refinery or steel mill) are a major element of the firm. Intangible assets play a core role for internet companies, and this makes financing difficult within the framework of a continental-style banking system with weakly developed stock markets. By contrast, the Anglo-saxon model of financing is more receptive for internet companies since this model´s emphasis on stock markets raises less obstacles for creating new firms than the banking-dominated system in continental Europe. This implies that the US and the UK should have first mover advantages in the internet business simply because of easier financing conditions. A drawback could be that both countries will have much more volatility in their national stock markets than previously. Continental EU countries have partly been able to remedy underdeveloped stock markets by launching new ones (New Market in the UK, Nouveau Marché in Paris, and Neuer Markt in Frankfurt, which is the most dynamic in Europe).

High technology, media and telecommunication were the leading sectors in 1999/2000 in Euroland, where they have strongly affected both stock and bond markets (BHF-BANK, 2000). Between October 1999 and March 2000 the Euro Stoxx Index increased by almost 50%. The subindex for construction increased by 11%, chemical industry by 6.5%, the automotive industry even faced stock price reductions. The stock market hausse was based on media, with gains of 185% and firms from the cluster telecommunications/technology/software with an increase of 150%. High rises of stock market prices in this area can be observed during certain periods, but a sharp fall in prices is also found at some point of time. In the three months after March 2000 the Neue Markt in Frankfurt (largely reflecting high technology, new economy companies, and biotechnology) lost about 40%. A similar volatility of sectoral stock prices can be observed in the US, the UK and Japan. The technology and new economy stocks often depend more on interest rate developments and less on cyclical factors than sectors in the traditional industry (except for chip producers which clearly depend on the business cycle). Young firms in technologically dynamic sectors face high fixed start-up costs and often have low revenues in a first expansion stage so that rising interest rates strongly undermine future discounted profits. High technology and software stocks are typically among the top ten stocks in most OECD countries since they are based on market capitalization telecommunication. High volatility of such stocks strongly affects the leading index. In the UK in 2000 it was even necessary to drop three high technology stocks from the index since strongly falling stock prices pointed suddenly to too low capitalization of the respective stocks. In most EU countries portfolio managers eager to shadow the leading index have complained that telecommunication firms have only a small percentage of shares traded since in many countries government still holds a large share in the former monopoly operator.

Many of the leading telecommunication firms have raised high loans to finance network expansion and the build-up of UMTS-mobile networks in Europe. Thus the traditional leaders in the corporate bonds markets, namely manufacturing industry and energy utilities, are no longer the biggest players. In 2000 this role has been assumed by telecommunication firms whose bonds fared well in the first quarter but suffered heavy price reductions in the second quarter. It was against this background that the FT (2000) ran: „European authorities fear banks have lent too much to help finance new mobile phone licenes. Regulators probe telecom loans“ as ist cover story on September 29, 2000.

4. Internet Economics and the Digital Divide within the OECD

There are serious impediments to the growth of electronic commerce in Europe. The low computer density in Europe is one major aspect, high costs of telecommunications and a fragmented EU market another one. National regulation of telecommunications might slowly lead to regulatory convergence via locational competition and benchmarking (cum „voice“): Locational competition means that foreign investors in telecommunications are likely to prefer countries with a rather pro-competitive approach. However, a serious caveat is the different sizes of EU countries. Large countries such as Germany, France, Italy, the UK, and Spain will always attract relatively high foreign direct investment, even if the respective national regulation is not strongly promoting competitiveness. Benchmarking means that operators active in various EU markets have the opportunity to compare regulatory approaches and then can publicly point out (this is the voice mechanism for obtaining regulatory adjustment) the best approach in the Community and identify crucial weaknesses in certain countries. Ie newcomers in telecommunications in the UK have complained in 2000 that the British approach of regulating local telephony is impairing competition, especially when compared to Germany where every newcomer can rent local access from the dominant operator at a regulated price. By comparing Germany and the UK the newcomers in the UK are able to identify weaknesses in British regulation. Similarly, internet service providers in Germany have pointed out in 2000 that the British regulator had forced the dominant telecom operator BT to offer a wholesale flat rate which is a prerequisite for functional competition in an internet market where internet service providers want to offer flat rates to households and firms. The British example (and the Dutch example) thus have helped to create pressure on the German regulator and Germany´s political system to reconsider the problem of internet pricing.

The industry association EUROBIT (1999) has identified various barriers which impair the growth of electronic commerce, where infrastructure bottlenecks are one important area:

„The growth of electronic commerce is restricted by high telecommunications costs, lack of availability of broadband communications, and lack of pan-European Internet backbone providers. Internet communications are generally established through telephone systems that were built to carry voice, not data, and thus have limited bandwidth, even with the fastest of today´s modems. Most consumers in Europe connect to electronic networks using ordinary telephone lines and pay local tariffs, which account, in many cases, for more than 60% of the cost of Internet access, a cost which US or Canadian consumers do not have. Research has shown that the use of Internet in the US took off when flat rates were introduced and consumers could calculate it as a fixed item on their household expenses.

Although the use of Internet in Europe has almost doubled within one year (1999 ISPO survey), the difference between the North, where deregulation is a fact of life, and South, where access costs remain too high, is still tremendous. 36.9% of private households are using the Internet in Sweden and 24.6% in Denmark, versus 3.9% in France and 6.1% in Italy.

Internet users throughout Europe require greater availability of competing telecommunications broadband services at more affordable prices. Telecommunications operators should widen their focus from voice telephony to meet the broadband needs of internet users. Today incumbant operators control the copper phone wires that comprise the „last mile“ of Incumbant Local Exchange Carriers´ (ILECs); central offices to end users and existing lines can not be reasonably reproduced by competitors. These must be unbundled. Emerging alternative providers concentrate their businesses in high-density areas where alternative infrastructure is available. There is little if any offer for broadband access in remote areas... Furthermore, Internet infrastructure in Europe, which is different in each country, is not co-ordinated across the EU, even at the major backbone level. It is much easier to do business over the Internet within the US or within one EU country. Corporate users pay more than 10times the price for a leased line in many countries in Europe than users in the US. Europe also has far fewer network access points. The Commission and national regulators should recognize that this is a significant cost barrier to carrying out pan-European business. Finally, the Commission should encourage the growth of pan-European Internet infrastructure providers, support local initiatives on developing Internet 2 architecture and platforms, and ensure full network interoperability across Europe.“

From this perspective the EU's problem of catching up with the US in terms of economic growth and New Economy dynamism is crucially related to pro-competitive national (and supranational) deregulation of telecommunications. While the Commission has made suggestions in 2000 to unbundle local access (in particular in favor of internet access) there is insufficient framework for regulation in the EU; the divergent approaches of national telecom regulators impair the emergence of a single market in telecommunications and internet in the EU.

The EU´s regulatory structure is particularly inconsistent in the field of 3G mobile licences, where several countries (ie UK, Germany, Netherlands, and Austria) have decided to award licences on the basis of competitive bidding, while other countries have adopted various forms of a beauty contest procedure (ie Spain with no licence fee, France with a modest licence fee, and Italy). This is not consistent with creating a level playing field in Europe since those companies which have obtained a UMTS licence in their home country at a discount fee enjoy a financial advantage when bidding in other markets. Spain´s Telefonica, Fance Telecom and Telecom Italia thus enjoy an unfair advantage over BT, Deutsche Telekom, KPN and AT).

5. Telecoms Modernization, Productivity and Growth

Is modern telecommunications and the internet a kind of new industrial revolution? In the perspective adopted here the answer is clearly yes since mobile and fixed internet will considerably change the mode of industrial organization in small and large firms and since it contributes to creating far bigger markets (in terms of geographical coverage) than ever vefore: global markets characterized by low transaction costs and strong competition. Transaction costs will fall and network effects (where users benefit as more users are linked to the network) will be exploited in the new economy.

Larger markets will enhance growth via economies of scale in certain sectors and more competitive markets will stimulate innovations worldwide as combining knowledge via telecommunication networks and superfast internet is becoming cheaper and easier. It is no coincidence that the growth rate in patenting as measured by applications at the European Patent Office is highest in telecommunications, where deregulation, privatization, and the dynamics of internet start-ups have strongly encouraged Schumpeterian dynamics in the late 1990s.

ROBERT GORDON (1999) has suggested that there is no productivity effect from the New Economy in the US. He points out that US growth rates of nonfarm private business output per hour in the three intervals extending between 1950:2, 1972:2, 1995:4, and 1999:1 were 2.63, 1.13, and 2.15 percent, so that recovery in productivity had been 1.02 percentage points in the late 1990s which is 68% of the 1.50 point slowdown between 1950-72 and 1972-95. The acceleration of productivity growth in the second half of the 1990s in his view can be explained by three factors: (1) improved methods of measuring price deflators, (2) the normal procyclical increase of productivity in periods like 1997-99, when output grows faster than trend, and (3) the explosion of output and productivity growth in durable goods, entirely due to the production of computers. According to GORDON in the US there has been no productivity growth acceleration in the 99 percent of the economy located outside the sector which produces computer hardware. Productivity acceleration in his view is mainly due to the computer manufacturing industry, which produced an enormous productivity growth since 1995. Not the use of computers, rather the improved production technology of computers would explain in GORDON's view the noncyclical productivity growth. He points out that BEA deflator for the final sales of computers in GDP declined at an annual rate of 36.7% per year in 1997:1-1999:1, which indicates, together with rising number of computers produced, a high productivity growth in that part of the economy which produces computer hardware.

GORDON (1999, p,.4) argues: „Because productivity growth in durable manufacturing between 1995:4 and 1999:1 has more than doubled from its 1992-95 trend...the portion of the productivity recovery remaining to be credited to the nonfarm nondurable (hereafter NFND) business sector is correspondingly less than when durables are included. The growth rates of NFND output per hour in the intervals cited above (1950-72, 1972-92, and 1995-99) are, respectively, 2.68, 0.80 and 1.50 percent. Thus the recovery in NFND is not 69% of the slowdown, as is the case for the total nonfarm business, but a much smaller 37% (0.70/1.88)." When he takes into account the productivity explosion in computers his analysis arrives at optimistic estimates of growth in potential GDP and „benign implications for the misnamed Social Security „crisis"." The GORDON productivity view is, however, doubtful. Rather one may doubt both the methodology of growth and productivity analysis used by GORDON.

GORDON fails, among other things, to take into account the increasingly important role of software expenditures, which could be viewed as similar to investment in equipment. If one adopts a broader investment concept which includes equipment, software, and innovation it is obvious that capital intensity in the civilian nonfarm sector has increased in the 1990s and that this should account for part of the structural labor productivity growth.

ROBERT GORDON´s general criticism is that expectations are too high with respect to the new economy, at least when judging on the basis of US productivity growth figures for the 1980s and 1990s. With slightly modified/updated data, GORDON argued in the FT (Financial Times, July 26, 2000, p.13) that the high annual US productivity growth of 2.75% in 1995-99, as compared to the 1.42% per year in 1972-95, does not reflect computer use and internet expansion, rather the differential of 1.33 points can largely be attributed to cyclical effects (boom situation), which account for ½ a percentage point, the effects of technological acceleration in computers and peripherals (0.29) and in other durables (0.28), and contribution of capital accumulation in computers („capital-deepening“: 0.33). In an opaque misunderstanding of the core of the ICT dynamics, ROBERT GORDON wrote in the FT: „Exponential increases in computer capability collide with a fixed supply of human time to invoke diminishing returns to computer capability at an unprecedented rate. In my case, the PC on my desk represents 30 times as much GDP as the one that sat there in 1983, but neither my mind nor fingers move any faster now...The need to protect market share against competitors explains much internet activity...Second, much internet content is not truly new, but rather consists of existing forms of information (ie airline schedules and stock quotes) made available more cheaply and conveniently. In contrast, the great inventions of the late 19th century created truly new products and activities...[There is] the problem of the growing distraction of personal web surfing. One research service found that people spend twice as much time online at the office as they do at home...The rapid onset of diminishing returns suggest that the greatest productivity contribution of computers lies in the past, not in the future.“

The first point mentioned by GORDON simply shows that he personally has not adjusted his mode of research over the past 25 years or so. Many of my colleagues and myself have used modern computers, improved global data access, and the fast internet to switch from isolated individual research to building national and international multi-person research networks in which each member of the team is more productive (produces higher value-added per hour) than if each researcher were to work in isolation for himself/herself. Prudent networking gives brains to the power 2+x, which is also the reason why large computer firms or pharmacutical companies have created flexible international research networks which show increasing innovation productivity: Innovation cycles are indeed shortening. Moreover, advanced software allows the researcher to no longer rely on typing with his/her finger, rather voice controlled computers will increasingly be used in the future. Again, the main productivity growth is not from computers per se but from a different innovative mode in which we use the increasingly powerful machines.

If, however, one wants to measure the impact of computers on growth, one should not simply focus on the production of computers, but on the use of computers, which is reflected by domestic computer production and the sectoral net export balance. The US and most EU countries have had high net imports of computers for many years, so that distinguishing between producing and using computers is quite important. GORDON is looking only at computers produced in the US, which is a very biassed view for assessing the new economy. An adequate specification of the production function would be:

(1) $Y_t = AK^{\beta 1} Z^{\beta 2} T^{\beta 3} L^{\beta 4} Q^{\beta 5}$

where K is capital (without computers), Z computers used, which is cumulated computers produced domestically plus the sectoral import-export balance (not GORDON's number of computers produced!), T use of telecommunications, L labor, and Q an index for the number of patents and accumulated imported licences; A is a scale variable. Based on a logarithmic econometric specification one should expect all elasticities to be positive. A simplified and modified approach which lumped K and Z together was estimated for Germany with significant positive parameters (JUNGMITTAG/WELFENS, 1998). The basic idea behind the inclusion of the use of telecommunications (ie proxied by the number of minutes used) is that the number of phone call units reflects the diffusion of new knowledge and best practice technologies.

At an early stage of introducing digital network technologies there are high adjustment costs. People have to be trained to learn new technologies and more people have to be hired in most firms for keeping complex computer systems running. All this implies that labor productivity growth in an economy with more powerful and networked computers is not as high as one might expect in the long run. It is also very unlikely that established and proven procedures in certain industries, say catalogue retailing, can, at low costs, be transfered into a digital internet universe. Only a few products can be digitized and sold directly via the net, and if the net is used in a straightford way intellectual property rights will be violated; ie think of illegal sales or trades of music or movies.

If people increasingly spent time in the office on personal surfing in the 1990s, as GORDON is suggesting, then there is little reason to complain, except from the legitimate (microeconomic) perspective of the firm, whose shareholders will not like personal websurfing of employees. However, in a macroeconomic perspective, we simply have an indication that labor productivity growth is even higher than the official statistics show in the 1990s for the US. Time spent in the firm for personal websurfing naturally should not be included in labor hours worked.

With the expansion of internet use in firms (US and UK around 50% in 2001, household user density around 30%), one should expect positive network effects of using computer and telecommunications networks. Clear economic analysis which takes into account the nature of network effects, will, unlike GORDON suggests, anticipate temporarily increasing returns of computer-internet-telecommunication application. The positive network effects will finally fade out as there are saturation effects on the one hand, and, on the other hand, networked societies face more complex technical disturbances (with internet virus problems just being a small part of it).

Careful analytical and empirical research will show the size and persistence of internet-related productivity acceleration for the US and Europe. Moreover, from a broader welfare perspective, one should not only ask about the internet's impact on output growth but also about wealth effects. If the internet raises investor market transparency (so that risk for investors is falling on average), reduces the cost of creating new firms, and increases profit rates, then this would also be beneficial. Indeed, one may assume that part of the US benefit of being a global leader in the internet results from rising foreign direct investment inflows which improve the terms of trade for the US.

The US perspective is partly summarized by a quotation from the Brookings Institute website (see BROOKINGS, 2000: www.brookings.edu/es/research/ ra12summary.htm): „The growing importance of the Internet revolution has become a well accepted fact, if not the symbol, of the new American culture. By one recent estimate, the so-called „Internet Producing„ sector of the economy – the companies responsible for hardware components and the content – currently account for over 300 billion in annual gross output, or a little over 4 percent of GDP. More broadly, according to the Department of Commerce, the entire „information-technology producing„ industry – a broader concept counting, among others, communications service – accounted for twice as much GDP in 1999...What is not widely understood – or even studied, for that matter – is how much the growth of the U.S. economy, and thus living standards of average Americans, are likely to benefit in the next 5-10 years, from the continued growth in usage of the Internet, both here and abroad."

6. Computer Lag, Network Fragmentation and Education Deficits: Policy Challenges in a European Perspective

The internet age offers new opportunities for growth, international cooperation, and democracy, but also raises major challenges. Growth-enhancing effects from the internet can hardly be strong if intellectual property rights are not strengthened in the field of digital services. Here we have a new challenge for the WTO, possibly also for the OECD. The internet brings about new opportunities for internationalization of business. For decades the leading nation in this field has been the US, followed by the UK, the Netherlands, Germany, France, Switzerland, Sweden, and Japan, at least if we judge on the basis of foreign direct investment outflows. The US is slightly better positioned than the EU when it comes to developing new global business on the basis of existing foreign investment positions; so far we have focused on FDI outflows, but the FDI inflow position could be equally important as foreign subsidiaries located in the US and Europe will pick up new internet developments observed on both sides of the Atlantic. Europe has both a problem and an advantage: The EU is highly fragmented in terms of political systems and languages, at the same time the diversity of languages could be an advantage when it comes to building electronic bridges into the world community.

If one takes a look at the business community on both sides of the atlantic one can identify certain asymmetries which are likely to tilt the balance of global e-commerce in favor of the US for decades to come. The starting point here is three observations:

Computers in the US are generally cheaper than in Europe, most notably than in Germany, which is the largest economy in Europe. A major reason is that EU markets are much more fragmented than the US market, despite the single market program of 1992; in Germany one additionally finds an artificial restriction of price competition in the retail market which is responsible for computer prices in Germany (compared with US figures including average rebates) to exceed US prices by 10-20%. This is a margin which will not fully mobilize transatlantic price arbitrage as computers differ in terms of keyboards and warranties. The implication is that the adoption rate of computers in the EU is lower than in the US, and this is a serious disadvantage in a period of increasing use of computers in a networked society. Consequently, the EU suffers from a lag in computer diffusion which impairs growth and job creation; the latter is true despite the fact that one might argue that more and faster computers can replace unskilled workers, the more important aspect is that more computers per capita imply higher labor productivity and hence a higher demand for skilled (and unskilled) workers.

In the US (see the diagram in appendix II) has had a long term fall of the relative price of ICT capital relative to non-ICT capital in the period 1960-2000. The growth rates of ICT capital formation have accelerated in the second half of the 1990s as the fall in the relative capital price has accelerated. However, growth rates of non-ICT capital also were increasing in the second half of the 1990s. Investment in software which is difficult to measure certainly also has been crucial for the US acceleration of productivity and growth in the 1990s.

Europe suffers from market fragmentation in the telecommunication sector. This is the consequence of decades of state-owned national network operators and the difficulties in developing EU wide operating companies. France has a law which requires government to maintain at least 50% of the national fixed line operator, Spain and Italy have golden shares which still allow the respective government to interfere with the business of dominant network operators despite full (Spain) or partial privatization. The European Commission can hardly allow mergers among the giants with the former national operator having about 90% market share after the first two or three years of post-1998 liberalization. Rather the giants are acquiring small newcomer firms and then trying to push for higher market share, but this in any case is a cumbersome way to develop EU-wide network operators. The EU is likely to even lose its leadership in mobile telephony within a few years, as US network operators offer uniform low mobile rates for the whole US. Compared to the US, mobile phone calls in the EU are expensive as soon as there is a cross-border call within the EU. This impairs market and internet integration in Europe compared to the US, to the extent that mobile internet use should become increasingly popular. In comparing the EU and the US in terms of ownership structure of telecommunication firms, one will find that EU companies very often own a minority stake in EU mobile newcomers, while EU firms are rarely active in the US (we disregard insignificant ventures here). As dominant network operators of country i in the EU is acquiring a share in mobile newcomers in neighboring markets the role of US firms is often strengthened.

The internet brings an acceleration in the diffusion and generation of knowledge. Therefore education, learning, and (re-)training have become much more important, and the ability of universities and schools in reacting to the new challenges is crucial. Here the mixed US university system clearly has an advantage as the competition between private and public universities on the one hand, and strong competition between private universities on the other contributes to a relatively high reaction speed of the US university system.

A major advantage of the US system is that it relies on a system of highly professional and competitive think tanks and help to shape responses to new challenges within months. There is nothing comparable in Europe, perhaps with the exception of the UK and its London-Cambridge-Oxford network.

It is particularly surprising that Germany does not have a broad network of globally competitive think tanks whose publications and websites would reflect intellectual leadership in Europe and worldwide. The most visible problem of the EU is that there is no genuine European network of think tanks which would openly and professionally discuss new issues and state alternative policy options so that policymakers in Europe could chose from a well-reflected policy menu. The US finally benefits from liberal immigration and active integration policies which will continue to attract many of the most talented people worldwide. Some of them in turn will contribute to global leadership of the US in education and distance learning. Videos and DVDs covering almost all fields of learning and university knowledge are already sold in the US, and professors from leading private US universities are among the most active developers of new internet-compatible material.

Given the strong initiatives of the European Commission in liberalizing EU network industries (telecommunication, energy), one can only hope that Brussels will actively try to foster a debate in Europe about how the EU can meet the global US internet leadership. Strong global competition between the US, the EU, and other key actors is in the interest of global economic welfare. Nobody in the EU should complain about US dominance and leadership in computers, telecommunications, and the internet. EU countries´ national governments must draw their respective conclusions. Strong competition in open markets, more private universities (with subsidiaries abroad/worldwide), and more venture capital for internet pioneers is needed in the EU. There is a transatlantic anomaly as long as most Europeans consider it natural that they often create their first companies in New York, Boston, San Francisco, or Los Angeles, while hardly any young American seriously considers EU countries as a natural choice for starting his/her first internet company. Finally, one should consider public initiatives to create a network of private think tanks in the EU.

Appendix I: Accounting for the US Productivity Acceleration in the 1990s

Tab. A13: Accounting for the Productivity Acceleration in the 1990s

[Private nonfarm business sector; average annual rates]

Item		1973 to 1995	1995 to 2000	Change (percentage points)
Labor productivity growth rate (percent)		1.39	3.01	1.63
Percentage point contributions:				
Less:	Business cycle effect	.00	.04	.04
Equals:	Structural labor productivity	1.39	2.97	1.58
Less:	Capital services	.70	1.09	.38
	Information capital services	.41	1.03	.62
	Other capital services	.30	.06	-.23
	Labor quality	.27	.27	.00
Equals:	Structural TFP	.40	1.59	1.19
Less:	Computer sector TFP	.18	.36	.18
Equals:	TFP excluding computer sector TFP	.22	1.22	1.00

Note.—Labor productivity is the average of income-and product-side measures of output per hour worked. Total factor productivity (TFP)is labor productivity less the contributions of capital services per hour (capital deepening)and labor quality.

Productivity for 2000 is inferred from the first three quarters.

Detail may not add to totals because of rounding.

Sources: COUNCIL OF ECONOMIC ADVISERS (2000); Department of Commerce (Bureau of Economic Analysis) for output and computer prices; Departmentof Labor (Bureau of Labor Statistics)for hours and for capital services and labor quality through 1998;and Council of Economic Advisers for the business cycle effect and for capital services and labor quality for 1999 and 2000.

Appendix II: Relative Price of ICT Capital to Price of Capital and Growth Rates of ICT Capital Formation and Non-ICT Capital Formation in the US

Fig. A14: Relative Price of ICT Capital to Price of Capital and Annual Growth Rates of ICT Capital Formation g(q(ICT)) and Non-ICT Capital Formation g(q(K-ICT)) in the US

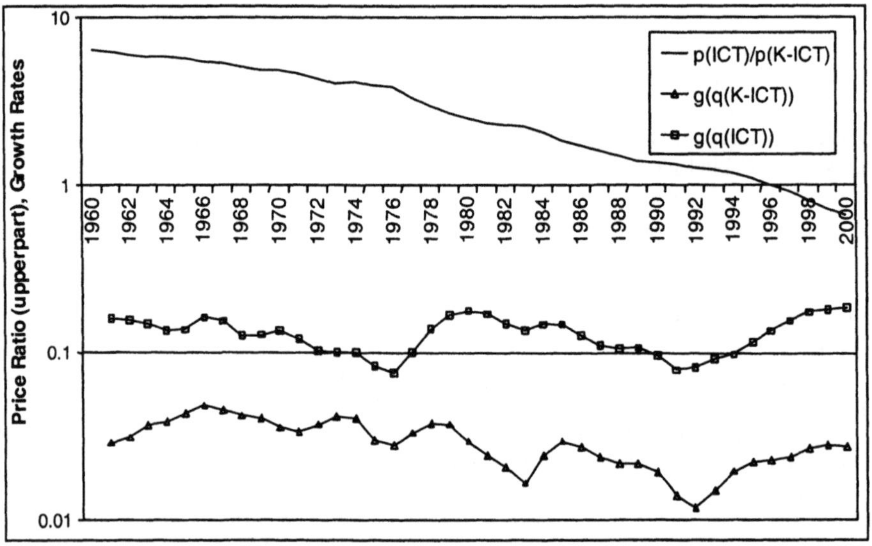

Source: BEA (www.bea.doc.gov), own calculations.

B. The Internet in Macroeconomic Analysis and Growth Analysis

1. Short Term and Long Term Aspects

From a theoretical perspective one may expect that the internet will affect economic development in many crucial ways. The internet reduces transaction costs in markets and therefore it should reinforce the role of markets; however, internet sales did not exceed 1% of overall sales in the US in 2000 which seems to show that there are problems with internet based transactions – security problems with payments being certainly the most crucial. Internet auctions have, however, basic advantages once an electronic marketplace with all major actors on the demand side and the supply side can be established: Walras is back to the real world in the sense that the internet is an ideal platform for fast auctioning; e-bay is a successful company which demonstrates this for the consumer business. Several industry platforms built for sourcing inputs cheaply and effectively, signal the same for part of the business community: The automotive industry and the chemical industry are two important examples. Airline ticketing via the internet is another example which is relevant for both consumers and the business sector. However, the internet also reduces firms internal transaction costs: If internal transactions costs should fall more than market transaction costs then the internet will stimulate the growth of larger – often multinational –companies. Indeed, both could occur at the same time, namely if in some sectors relative market transaction costs fall while in other sectors relative transactions costs within firms are reduced. Reality will tell which effect will dominate. These changes will affect the structure of the economy and adjustment – including adjustment speed – in markets.

Another aspect is arbitrage which has national and international aspects. The internet could raise the price elasticity of demand in national markets since in the case of tradable goods and (digital) services consumers have new alternatives to substitute between suppliers across countries. From a macroeconomic perspective this could have considerable effects with respect to the exchange rate variable and exchange rate policies where demand elasticities (Marshall-Lerner condition) and supply elasticities (Robinson condition which takes into account both supply and demand elasticities) are known to play a crucial role.

The internet also might reinforce the law of one price internationally, namely to the extent that it reinforces arbitrage internationally. What can we say about the empirical evidence: Empirical evidence (EUROPEAN COMMISSION, 2001, p.2) suggests that internet retailers charge lower prices than conventional retailers partly as a result of reduced retailing costs. Initially the use of the internet seems to lead to an increase in price dispersion since only a rather small share of consumers buy the product at the lower prices available on the internet. Beyond this short term effect there is the observation that as the share of people using the internet becomes more widespread price dispersion falls again. A rather surprising feature of the internet is that price dispersion on the net has not been found to be lower than for conventional retailers. One plausible explanation is that the internet offers greater scope for product differentiation. In 2000 e-commerce represented only about 0.5% of GDP of the Community. With e-commerce becoming more important in the long run it remains to be seen what the long term effect of the internet on price dispersion will be.

2. The Internet in Macroeconomic Analysis

Basic Macroeconomic Perspective

The internet has microeconomic relevance in the sense that it might affect individual behavior; eg it could accelerate diffusion of behavioral changes. From an economic perspective the more important issues are in the macroeconomic arena. From a macroeconomic perspective the internet has seven potential key effects:

It will reduce capital intensity due to two effects: By facilitating intra-company coordination across the value-added chain it will reduce inventory-out ratios. At the same time it will encourage the substitution of real capital by digital services and software, respectively. According to figures of the OECD (2000) the share of software-expenditures to GDP reached almost 3% in the mid-1990s, putting the US in a clear global leader position in the field of software usage.

To the extent that software is complementary to computer equipment the increasing use of software will raise capital productivity which in turn will raise the wage premium for skilled workers being complementary to real capital and software.

It will affect structural change, eg by creating new opportunities for outsourcing which in turn will affect the demand for money and economic competition. This could have direct and indirect effects on the real and nominal interest rate.

The internet will stimulate diffusion and therefore may stimulate innovation. The internets role in the creation of innovative digital services, will be linked to an increase in trade and, according to one strand of the New Growth Theory – emphasizing the role of rising number of product varieties –expected higher sustained growth.

It reduces international trading costs and therefore will raise export-GDP ratios (and import-GDP ratios). This may be accompanied with faster international diffusion of technological progress – embodied in technological-intensive products and in new digital services.

The internet reinforces global transparency with respect to information about competing investment locations. This will reinforce the quest for mobile capital; since the increase in global price transparency will stimulate firms in OECD countries to spezialize more on innovative goods and services – allowing them to fetch higher margins in markets – ownership specific advantages will become more important and therefore foreign direct investment will continue to increase strongly. Improved global transparency in regards to investment conditions will provide, in particular, improved information about regions in OECD countries which may reinforce concentration of FDI flows among OECD countries (here we ignore China as a special case of a rapidly growing large country in Asia). This may exacerbate the North-South divide, already expected to expand due to the fact that roughly 50% of the adult population in developing countries are illiterate and thus lack an important complementary skill with respect to the internet.

The internet is increasing price transparency and creates larger – often international – markets for tradable goods and services; this implies rising opportunities for substitution and therefore declining price elasticity. Inflationary pressures may thus be smaller in the New (International) Economy than previously. The short-term inflation-unemployment trade off may be affected – the NAIRU will fall. The NAIRU has indeed fallen in the US in the 1990s. However, one may also assume that profit rates in sectors where firms used to benefit from price differentiation, will fall as price differentiation will become more difficult. With a linear demand curve $p= a-bq$ and a flat marginal cost curve $k'=c_0$, perfect price differentiation (an extreme case in a world without the internet) will lead to an average sales revenue of $(a-c_0)/2$, while in a world of perfect competition – with a resurrection of the Walras auctioneer in the internet – will lead to a uniform price of c_0,. For the average firm the marginal value product of labor will therefore fall with a switch to the internet so that there will be a once-and-for-all reduction in employment; this could raise the natural rate of unemployment, however, we will see in the next paragraph that the long term rate of unemployment may indeed decline for two reasons.

Phillips Curve and the New Economy

The original Phillips curve established a relation between the nominal wage rate - which will be equal to the sum of the inflation rate (π), productivity growth – and the unemployment rate u; the relation between inflation and unemployment is nonlinear (f(u)). According to the Friedman critique the short term Phillips curve must be augmented by the expected inflation rate, and if this rate is equal to the actual inflation rate – in the long run – then the unemployment rate will be determined by the short term trade-off parameter in the function f(u) and the growth rate of labor productivity (a).

(I) $\pi = f(u) - a + \pi^E$

Let us assume a very simple specification of f(u), namely v/u. We then get as the natural rate of unemployment, u# from the equation:

(I.I) $v/u= a$

(I.II) $u\#=v/a$

If the short-term inflation-unemployment trade off coefficient v should fall in the New Economy, that is in the networked internet society, one may anticipate a fall in the long term unemployment rate (NAIRU). Moreover, if the internet expansion raises the growth rate of labor productivity growth (a), the long run unemployment rate should also fall.

A rather interesting variant of the Phillips curve discussion is from MANKIW (2001) who derives a New Keynesian-Phillips-Curve from a general equilibrium model. There are two types of rigidities, namely nominal rigidities and real rigidities. Nominal rigidities are related to the frequency of price adjustment (λ), real rigidity refers to the parameter α, namely to the reaction of economic activity to changes of relative prices. Small parameters α and λ stand for high rigidities; rigidity explains why inflation is not reacting to a high unemployment rate (u) in the short run; the natural rate of unemployment is u_{nat}. We thus have a Phillips curve which allows temporary deviations of the unemployment rate from the natural rate as long as actual inflation is not equal to the expected inflation rate.

(*i*) $\pi_{t\cdot} = E_t\, \pi_{t+1} - [\alpha\, \lambda^2/(1-\lambda](u_t - u_{nat})$

An interesting aspect of the internet is that it reduces price adjustment costs so that the frequency of prices changes of standardized products will tend to increase. At the same time the internet stimulates diffusion and therefore stimulates innovation and the R&D-GDP ratio, respectively. This, however, will reduce price elasticity of demand, moreover, with a rising share of innovative goods the notion of quality goods becomes more important, and for those frequent price changes are rather uncommon. This should limit the rise of the frequency of price adjustment in the medium term. One also may note that a rising share of R&D costs in overall costs implies a higher share of sunk costs which might raise price flexibility if newcomers attack incumbant firms – they could ignore the (increased) share of sunk costs in their strategic pricing decisions which aim at fending off entry of newcomers. This creates some ambiguity with respect to the link between the internet and the Phillips curve. An open question also is whether real rigidity will reduce as a consequence of the growth of the internet. Here only empirical research can clarify the issues.

Macroeconomic Perspective

Considering medium term effects of the internet, we are interested in analyzing how the internet – proxied eg by internet host density or user density (Z) – will affect the economy. We will use a modified macro model in which consumption C depends on disposable income $Y(1-\tau)$ and product varieties as proxied by Z. Government real expenditure G is exogenous. Investment I depends on the real interest rate r, Z and the expected nominal exchange rate e^E. Net exports depend positively on Y* (foreign variables carry a star) and the exchange rate e, negatively on Y; Z also plays a role. The real money demand m depends on output, the interest rate, Z and the expected exchange rate. Our model setting is rather Keynesian in the sense that we have assumed unemployment in the initial situation and sticky price levels

P at home and abroad. Net capital imports Q are assumed to depend on the nominal interest rate i, and the foreign interest rate i*, the tax rate τ, the internet variable Z and the expected exchange rate.

We assume that the internet will have a positive influence on consumption since the internet brings digital product innovations which should stimulate consumer demand. At the same time we assume that the internet is raising investment as expected rates of return are increasing. Moreover, we assume that net exports will positively react to the internet; this is plausible for a country leading in internet applications: digital product innovations and cutting trading costs should indeed stimulate net exports X'. The impact on the demand for money is rather unclear. One may assume that the use of the internet will lead to a reduction in the demand for money (M1) since the use of the internet accelerates payments. At the same time we cannot overlook the fact that the internet stimulates outsourcing, thus the demand for money in the business sector could increase. In addition, the demand for money could raise turnover figures in stock markets which would increase the demand for money; so the partial derivative of the money demand with respect to the internet is ambiguous. Finally, we assume that net capital imports Q will increase with higher Z for the type of country considered, namely a country leading in internet applications (eg. US, UK, Finland, Sweden). This leads to the following system of equations where we set i=r due to the absence of inflation:

$$(1) \quad Y = C(\underset{+}{Y(1-\tau)}, \underset{+}{Z}) + G + I(\underset{-}{r}, \underset{+}{Z}, \underset{+}{e^E}) + X'(\underset{-}{Y}, \underset{+}{Y^*}, \underset{+}{Z}, \underset{+}{e})$$

$$(2) \quad M/P = m(\underset{+}{Y}, \underset{-}{i}, \underset{?}{Z}, \underset{-}{e^E})$$

$$(3) \quad Q(\underset{+}{i}, \underset{-}{i^*}, \underset{-}{\tau}, \underset{+}{Z}, \underset{-}{e^E}) = [-X'(\underset{-}{Y}, \underset{+}{Y^*}, \underset{+}{Z}, \underset{+}{e})]$$

Differentiating equations (1) to (3) we get:

$$\begin{pmatrix} 1-c_y(1-\tau)-X'_Y & -X'_e & -I_r \\ m_Y & 0 & m_r \\ X'_Y & X'_e & Q_i \end{pmatrix} \begin{pmatrix} dY \\ de \\ dr \end{pmatrix} = \begin{pmatrix} 0 & 0 & 1 & I_{eE} & 0 & X'_{Y^*} & -C_y Y & C_Z+I_Z+X'_Z \\ 1 & -M & 0 & -m_{eE} & 0 & 0 & 0 & -m_Z \\ 0 & -Q_i & 0 & -Q_{eE} & -Q_{i^*} & -X'_{Y^*} & -Q_\tau & -X'_Z-Q_Z \end{pmatrix} \begin{pmatrix} dM \\ dP \\ dG \\ de^E \\ di^* \\ dY^* \\ d\tau \\ dZ \end{pmatrix}$$

$$\mathbf{A} \qquad \mathbf{x} \quad = \qquad \mathbf{B} \qquad \mathbf{y}$$

The determinant of matrix $|\mathbf{A}|$ is:

$$|\mathbf{A}| = -(1\text{-}c_y(1\text{-}\tau)\text{-}X'_Y)\,X'_e\,m_r + X'_e(m_YQ_i - X'_Ym_r) - I_rm_YX'_e.$$

If m_YQ_i is high in absolute terms (ZZ curve is flat because $|Q_i|$ is high) we have $|\mathbf{A}|>0$. We therefore get – by using Cramer's rule (with C_Y as the partial derivative of C with respect to disposable income $Y(1\text{-}\tau)$, and similarly denoting other partial derivatives) the following results; the macroeconomic analysis shows the following results if we consider Z as an exogenous variable which is technologically determined:

(1) $de/dZ = |\mathbf{A}|^{-1}\,[(1 - C_Y(1-\tau) - X'_Y)(-m_ZQ_i + (X'_Z + Q_Z)m_r)$
$-(C_Z + I_Z + X'_Z)(m_YQ_i - X'_Ym_r)$
$-I_r(m_Y(-X'_Z - Q_Z) + X'_Ym_Z)]$,

therefore

$de/dZ < 0$ if $m_Z \geq 0$ and $m_YQ_i > X'_Ym_r$ (other cases are possible);

(2) $dr/dZ = |\mathbf{A}|^{-1}\,X'_e\,[(1 - C_Y(1-\tau))m_Z + m_Y(C_Z + I_Z - Q_Z)]$,

therefore

$dr/dZ > 0$ if $m_Z \geq 0$ and $C_Z + I_Z > Q_Z$
$dr/dZ < 0$ if $m_Z \leq 0$ and $C_Z + I_Z < Q_Z$.

Only empirical analysis can fully clarify the short-run impact of the internet on the exchange rate and the real interest rate. As it seems plausible that $m_Z \leq 0$ and $C_Z + I_Z < Q_Z$ we suggest that the internet will reduce the real interest rate. Such an effect could indeed be one element of a "New Economy". A fall of r will in turn stimulate investment – including investment in ICT – and cause positive wealth effects while reducing budget problems. As regards the exchange rate impact in a system of flexible rates we consider an appreciation as plausible. This would concur with some of the US developments in the 1990s.

In a short-term analysis the effects of the internet on output are given by:

$$dY/dZ = |\mathbf{A}|^{-1}\,(C_Z + I_Z - Q_Z)(-X'_e)m_r + m_ZQ_i\,(I_r - X'_e);$$

and hence with the assumption that the system determinant be positive due to a sufficiently large interest responsiveness (Q_i) of net capital inflows: $|\mathbf{A}| = -(1\text{-}c_y(1\text{-}\tau)\text{-}X'_Y)\,X'_e\,m_r + X'_e(m_YQ_i - X'_Ym_r) - I_rm_YX'_e > 0$:

(4a) $dY/dZ > 0$ if $Q_Z < C_Z + I_Z$ and $m_Z < 0$,

(4b) $dY/dZ < 0$ if $Q_Z > C_Z + I_Z$ and $m_Z > 0$ (other cases are possible).

The case (4a) is quite plausible. We consider a negative impact of the internet on the demand for money as quite likely. Likewise, we consider it plausible that the impact of the internet on consumption and investment is much more powerful than the reaction of net capital imports with respect to the internet variable. Clearly, there are several new empirical questions which will have to be studied.

From this perspective of an analysis for an underemployed economy we clearly have a positive impact of Z on output – and employment. The international expansion thus leads us to expect a temporary increase in output and employment. At the bottom line the internet is like a positive supply shock.

From a broader perspective there are – beyond the short-term macroeconomic analysis – three other analytical challenges. A first challenge considers the interdependency between medium term output development and the expansion of Z, that is, there should be a medium term model which endogenizes the internet expansion. A second aspect not covered here in detail, concerns adjustment speeds in goods markets, factor markets and financial markets; as regards the latter the internet raises further the speed of adjustment. To the extent that there is a growing disparity between adjustment speeds in goods markets (plus labor markets) which are slow and financial markets which are relatively fast and indeed getting ever faster there is an additional potential for overshooting – or magnification effects –well known from the *Dornbusch model* in which the interaction of sticky output prices, fast interest rates and exchange rate movements cause instability and overshooting.

The third challenge will be treated subsequently: economic growth and the internet: This rather long term challenge concerns the issue as to how to incorporate the internet directly or indirectly into growth models. We will not look into the medium term problems, rather, we will take a closer look at growth modeling in the following section and present some new ideas.

3. Internet, Telecommunications and Economic Growth

The dynamics of the digital economy have to be put into perspective through a focus on growth theory and innovation analysis. Since we assume that the internet and modern digital telecommunications reinforce communication links and accelerate diffusion of knowledge, of particular interest are those studies which look at the role of knowledge, spillovers and product differentiation. There is clear evidence that R&D accumulation, human capital formation and trade positively contribute to economic growth (survey DURLAUF/QUAH, 1998).

GROSSMAN/HELPMAN (1991) show that economic opening up can raise output if there are external effects of knowledge and increasing returns for the production of tradables; moreover, under certain assumptions the rate of innovation can affect growth rates of output. Some empirical studies – which are in line with evolutionary analysis of Schumpeterian dynamics – have emphasized the role of product differentiation and product innovations for growth (FAGERBERG,

1988; SAVIOTTI, 1991). Accumulated knowledge – including tacit knowledge – plays a crucial role for technological progress. While codified knowledge is easily transferable across borders – at least within a region of countries with a comparable adaptation know-how – the international diffusion of tacit knowledge is rather slow and typically involves investment of MNCs abroad and skilled personnel working abroad. Empirical evidence suggests that the ability to innovate and emulate new technologies differs considerably across regions and countries (e.g. DOSI/PAVITT/SOETE (1990); GEHRKE/GRUPP (1994)).

The New Growth Theory has emphasized the fact that individual firms′ investment in R&D capital can have positive external effects which partly could be due to leakage effects in various forms, but also to network externalities. In the literature there are many empirical studies with a focus on individual countries, eg the US or France and Germany. PATEL/SOETE (1988) presented empirical analysis which puts the output elasticity of R&D at 0.21. BÖNTE (1998) suggested that elasticities for the R&D stock of selected sectors of West German manufacturing were in the range of 0.03 and 0.04, however, technology-intensive sectors were not earning above-average rates of return which is rather surprising. JUNGMITTAG/WELFENS (1998) estimated an output elasticity of the real patent stock of 0.23. MAIRESSE/HALL (1996) present results for France with elasticities in a range of 0-0.17. PATEL/SOETE (1988) suggest output elasticities 0.13 at the aggregate level for the R&D capital stock in France which is close to the results presented in COE/MOGHADAM (1993). With the (aggregate) output elasticity of R&D capital being relatively high the old problem of divergence between private and social rates of return reemerges. According to JONES/WILLIAMS (1997) the optimum amount of R&D exceeds private R&D by a factor of almost four in the US. As regards international spillovers the results by COE/HELPMAN (1995) and BAYOUMI/HELPMAN (1999) – the latter with focus on a broad range of countries – are important since they show that import-weighted foreign R&D stocks are important for total factor productivity growth

The internet will raise the degree of economic openness as many services become tradable due to the internet; moreover, the internet generally cuts international information and trading costs which should stimulate trade in goods and services generally. Moreover, the marginal costs of producing digital services often are very small (however, marketing costs could be considerable at the margin and might well differ across countries). The internet widens the window of opportunity for telecommunications operators which also benefit from innovative mobile technologies and applications making the internet available worldwide. It seems that the internet is a general purpose technology which will cut across all sectors: it will raise factor productivity in most sectors in the long run and it creates new opportunities for vertical or diagonal linking of skill-intensive and technology-intensive sectors. The obvious fact that the internet accelerates international diffusion of new knowledge could imply that international spillovers from R&D are increasing in the long run, the general acceleration in the diffusion of new knowledge creates an incentive for firms to become more innovative on the one hand, at the same time to become more international since recovering R&D

costs in larger (international) markets is a natural option to be pursued in an environment of intensified Schumpeterian dynamics.

The New Growth Theory has emphasized the role of spillovers from investment of individual firms, the growth-enhancing aspect of a growing number of product varieties (consumer products or intermediate products), learning-by-doing and other dynamic technology effects which naturally focus attention on human capital, R&D and technology-intensive trade (ARROW, 1962; COE/HELPMAN, 1995; GROSSMAN/HELPMAN, 1991, LUCAS, 1988; ROMER, 1986, 1990). Taking into account the role of telecommunications and the internet we will focus on some new aspects related to traditional and new growth theory:

- One aspect directly points to the role of the telecommunications network as the basis for accelerated diffusion of knowledge – an idea empirically investigated in JUNGMITTAG/WELFENS (1998; 2002) who found positive empirical evidence within Germany. The approach was based on growth accounting in the sense that output (without the housing sector) was explained in terms of capital input, labor input, patents plus real expenditure on license imports and the use of telecommunications as measured in minutes. Given our specification and the finding that all coefficients were significant and positive we can conclude that the use of telecommunications contributes to growth. This has interesting implications for the internet: Internet technology is a new communication platform which no longer relies on a dedicated connection between the sender of a message and the recipient, rather it uses a flexible routing. From a theoretical perspective we may anticipate a similar impact of the internet on growth as with the use of telecommunications. However, the use of time series is – for some years to come – impossible as time series with internet user data are very short.
- Communication can stimulate international trade. The internet can influence trade volume and the structure of trade (FREUND/WEINHOLD, 2000) positively so that there is a growth-enhancing effect, provided that trade contributes to output growth – this indeed is shown in the New Growth Theory. Moreover, international telecommunications contributes to trade as can be shown in a modified gravity model of trade (WELFENS/JUNGMITTAG, 2002).
- The internet sector itself can be a major field of innovation which, however, has certain new traits, including a multi-layer architecture which makes combining complex knowledge of large companies and new knowledge of many "partner firms" possible as long as there is interconnectivity. The internet has a high degree of such interconnectivity as computers with different operating systems can communicate with each other. The issue as to whether the internet sector – as one element of the ICT sector – requires a special approach in R&D policy is rather new. Countries with highly flexible labor markets may have better opportunities to expand in some layers of the internet sectors, countries with less labor market flexibility may have an advantage in strengthening other layers of the internet. Analysis of the OECD has shown that labor market flexibility and the ability to reorganize work relations swiftly at the firm level, are indeed requirements

to allowing a full exploitation some of the most crucial benefits of the internet (ARNAL/OK/TORRES, 2001).

- Since telecommunications has been regulated for decades in OECD countries and local telephony is so crucial for internet access, it is essential to take a closer look at the regulations of telecommunications; telecommunications has two specific traits which justify regulations – sector-specific rules for competition – for the sake of competition: There are falling marginal and average costs in the local loop; and there are general network effects in telecommunications so that expansion of the network and of value-added services will be easy to achieve if there is nondiscrimination in interconnection and if interconnection takes place at a price which is incentive-compatible and efficiency-enhancing. With new technologies emerging in mobile telephony there are new perspectives for substituting local access over the fixed network to some extent, although bandwidth differs considerably between the mobile networks and fixed networks – possibly until UMTS (3G) technology becomes generally available in OECD countries.

4. Internet, Innovativeness and the Real Exchange Rate

Our basic hypothesis is that the internet will stimulate R&D and innovation,– at least in the countries that are leaders in ICT expansion: the US, Sweden, Finland, Korea, UK and possibly Ireland. ICT leaders should be able to launch a higher rate of product innovations (including some which will combine elements of the Old Economy and innovative digital services) which will stimulate net capital inflows, including long term foreign direct investment. Countries in which the Old Economy is still dominating in the innovation process could face long term problems – in particular if they fail to specialize quickly in innovative products with relatively low price elasticity. In Europe, the problem of a dominant Old Economy still shaping the dynamics of the economic system, holds true for Germany and Italy which together account for roughly 50% of "Euroland"; this implies that Euroland might face a long term real devaluation as the following simple model will show. In this model Euroland is the home country (country I) and the US as the ICT leader is the foreign country (country II).

We assume that the change in the real exchange rate ($q=P/(eP^*)$) of Euroland depends positively on net exports x^{net} – or cumulated net exports as in RÖGER, 2002 - and negatively on relative innovativeness a^* in the rest of the world/the US: We therefore have the following differential equation (with positive parameters b and a^*):

$$dq/dt = bx^{net} - a^*q \quad (2)$$

We adopt the standard assumption that net exports are negatively dependent on the real exchange rate:

(3) $x^{net} = q^{\eta}$ (with $\eta<0$)

We then get the differential equation in q(t):

(2.1) $dq/dt = [bq^{\eta}] - a^*q$

(4) $q(t) = \{C_o e^{-a^*(1-\eta)t} + b/a^*\}^{1/(1-\eta)}$

The real exchange rate will converge and has the following steady state value :

(5) $q\# = (b/a^*)^{1/(1-\eta)}$

In the following graphical analysis we show the squared expression on the right hand side of equation (2.1) as the BB-curve which has a negative slope. The second expression of the right hand side of equation (2.1) is shown as the AA-curve. The intersection determines the steady-state value of q, that is q#.

Fig. B1: Model of the Real Exchange rate ($dq/dt=bq^{\eta} - a^*q$)

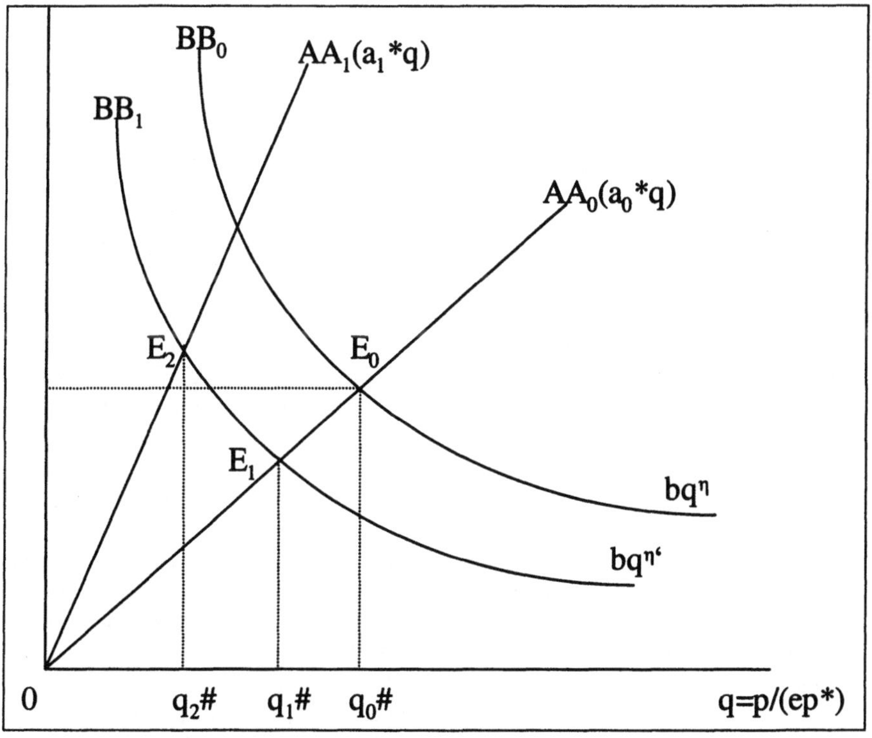

If the ICT lead of the US amounts to raising (once and for all) the innovativeness of the US this means that a* has increased. If Euroland at the same time should face more competition for its more standardized industrial products it will face an absolutely higher price elasticity in export demand. Euroland´s real exchange rate thus moves from a value equivalent to point E_0 to E_1 and finally to E_2: The US lead in ICT dynamics implies a real depreciation of the Euro. This has indeed been witnessed in 1999-2001, and this development will continue unless Euroland´s core countries catch-up with the US. This presents a formidable challenge since computer density in Italy in 1999 was only roughly ½ of the US figure, France had 2/3 and Germany about ¾. Internet user density and internet host densities were also much lower in Germany, France and Italy than in the US. The latter problem is partly related to the fact that there is no cheap internet flat rate available in Germany, France or Italy – much in contrast to the US and the UK (or Spain).

5. Information & Communication Technology and Growth: Some Neglected Dynamic Aspects in Open Digital Economies

5.1 Introduction

The OECD countries have witnessed considerable differentiation in economic growth in the 1980s and 1990s on the one hand, on the other hand the rate of innovation – as measured by patent applications – has increased in the US and Europe in the 1990s (COUNCIL OF ECONOMIC ADVISERS, 2000; WELFENS/AUDRETSCH/ADDISON/GRIES/GRUPP, 1999). The US has achieved a considerable lead in economic growth vis-à-vis Europe in the 1990s which was characterized by an unusual increase in labor productivity in second half of the economic cycle in the US (COUNCIL OF ECONOMIC ADVISERS, 2001). US labor productivity growth in the period 1973-1995 was 1.4% p.a., but in 1995-2000 it reached 3.1%. It is unclear what the reasons for the robust US productivity growth is. The Council of Economic Advisers has argued that falling computer prices and rising computer expenditures of wholesale trade, banking and the ICT sector itself play a crucial role for US growth in the 1990s.

The growth of the information and communication technology sector (ICT) in the 1990s has considerably contributed to growth in the US, especially in the second half of the 1990s. According to estimates by the US Department of Commerce (see the study DIGITAL ECONOMY) the ICT sector represented 8.3% of US output in 2000, but the sector contributed almost 1/3 of real GDP growth in the period 1995-99. Even more impressive is the contribution of ICT to investment in the US; it is noteworthy that computer prices have fallen strongly in the 1990s. In 1999 business spending for ICT equipment and software represented more than ¾ of the 12 percent real growth in total equipment and software spending. The contribution of the ICT sector in percentage points increased from 0.8% in 1994 when overall US growth was 4.2% to 1.6% in 1999 when the overall growth rate was 5%. In Europe only Finland and Sweden, possibly also Ireland, have an ICT sector which directly contributes significantly to economic growth. In 1989 Swedish Ericsson stood for about ½ percent of total Swedish GDP, but in 1999 the company stood for 2.6% of GDP and contributed 0.5 percentage points to economic growth. In Finland Nokia stood for 4% of GDP – and 1/5 of Finnish exports - and contributed a full percentage point to growth in 1999 (EUROPEAN COMMISSION, 2000).

The ICT sector (as defined by EITO, 2000) consists of IT-spending which reached 2.7% of GDP in the EU in 1999, but 4.5% in the US. Telecommunications accounted for 2.8% of 3.1% of GDP in the EU in 1999, for 2.8% in the US. Assuming that prices in the EU for telecommunication equipment and services are slightly higher than the US one may argue that the transatlantic differential in ICT spending is almost zero in telecommunications, but that the EU is far behind in the IT sector. According to OECD (2000) data which focus on the share of output and employment in ICT and on ICT exports plus ICT expenditures on research and development Germany is only in the lower third of the 29 OECD countries.

The use of telecommunications has contributed considerably to economic growth in the 1970s and 1980s in Germany, and the expansion of the internet could lead to a similar effect if it contributes to accelerating diffusion of information and knowledge, respectively (WELFENS/JUNGMITTAG, 1998, 2000; 2001; WELFENS, 2001). Network effects play a role both in the use of telecommunications and PCs (the internet).

As regards the US GORDON (1999) has argued that US growth acceleration in the second half of the 1990s – ignoring the cyclical effects – can fully be explained by the growth of output and productivity growth, respectively, in the computer producing sector and in the output of other durables. GORDON argues that there were no spillover effects in productivity to other sectors. GORDON argues that falling computer prices signal a falling marginal product of computer equipment. Subsequently, we will raise some doubts about this view. We also will look into the problem of modeling certain spillovers in a straightforward way. Indeed, we will address several aspects of the new economy and emphasize the potential role of spillover effects and network effects. We also consider the GORDON (2000) argument about the limited role of computer expenditures in the US as flawed: He argued that the expenditure on computers as a share of non-farm private business has stagnated at 1.3% in the late 1990s, however, he completely overlooks that the US is the leading country in terms of software expenditures which accounted for 2.7% of GDP in 1995 – for comparison France 0.9% (OECD, 1998). Analyzing the role of computers – and semiconductors - without taking into account software expenditures obviously is inconsistent.

5.2 The Basic Analytics of ICT

Analyzing the ICT sector means to look into technological dynamics, output-GDP shares and expenditures-GDP shares. Let us start with the simple observation that ICT is a highly dynamic sector of the economy. Judging by figures from the European Patent Agency the telecommunication sector has become that sector which has the highest growth rate of all patent groups; the growth rate in the period 1989-97 was 13.6% p.a., the growth rate of advanced electronics was 6.4% which made it No. 7 in the field of the ten fastest growing technology fields. This means that ICT is rather technology intensive and has some potential for technology spillovers (in the sense of the New Growth Theory à la Romer or else). The US plus Canada, the UK, Sweden and the Netherlands are positively specialized in telecommunications; they should strongly benefit from the global telecommunications boom which, however, will face some saturation problems in the long term when almost everybody has a mobile phone. In advanced electronics we find the US plus Canada and the Netherlands again, but also and strongly Japan (Tab. B1).

Tab. B1: Specialization (Relative Patent Share in Interval -100, +100) in 1995-99 in Technology-intensive Fields with High Growth Rates in Patents*

	Growth	USA	Japan	Germany	France	UK	Switzerland	Canada	Sweden	Italy	Netherlands
Telecommunication	13,6	10	-3	-34	-7	17	-75	50	70	-67	18
Turbines	10,6	-8	-74	-40	87	8	83	-84	-7	-96	-52
Railway Systems	8,5	-74	-41	67	9	-67	58	-22	-19	0	-26
Paper-Making Equipment	7,6	-4	-88	28	-71	-43	-54	30	85	-41	-62
Automobiles	6,7	-47	-14	57	35	-31	-84	-71	-12	10	-56
Medi. Sector, Instruments	6,6	46	-80	-38	-36	-7	38	-64	32	-20	-29
Advanced Electronics	6,4	-18	46	1	-21	-18	-52	42	-52	-44	48
Power Distribution	6,4	-20	8	16	34	-23	-27	-53	13	-7	-36
Agrochemicals	6,1	35	-59	0	-3	5	22	52	-53	-13	-69
Medi. Sector, Electronics	5,8	42	-31	-47	-64	-9	-48	-19	10	-65	35

* Average Annual Growth of Patent Applications at the European Patent Agency in 1989-1997

Source: FhG-ISI, Karlsruhe, preliminary

According to these figures the US, Canada and the Netherlands stand to gain particularly from the ICT boom worldwide. Japan is likely to benefit mainly on the electronics side. In the New Economy, there is, however, one caveat with respect patent figures; many internet-based services cannot be patented – copyrights and trade marks play an important role here.

Given the broad market opening up and deregulation of telecommunications which have taken place in continental countries after 1998 one may anticipate that telecommunications will remain a leading innovation field in the near future. Telecommunication operators drive for new services and technical sophistication in the age of the digital economy in turn is likely to stimulated advanced electronics. Innovation dynamics in telecommunications also might be accelerated by the ongoing internationalization of the telecommunication business; falling prices in voice telephony will stimulate creation of innovative services which can fetch higher prices in markets. What about falling computer prices as asignal of a falling marginal product of PCs? This view might be true if everybody uses ever more sophisticated computers only as a kind of typewriting device – certainly an exception in the US and Europe. With computer prices falling (k'_1 instead of k'_0) the price will fall – with a given demand curve – from p_0 to p_1; if this goes along with network effects so that demand curve DD_1 holds instead of DD_0 the price still falls to p_1 (see Fig. B2).

Fig. B2: Technological Progress, Prices and Marginal Utility

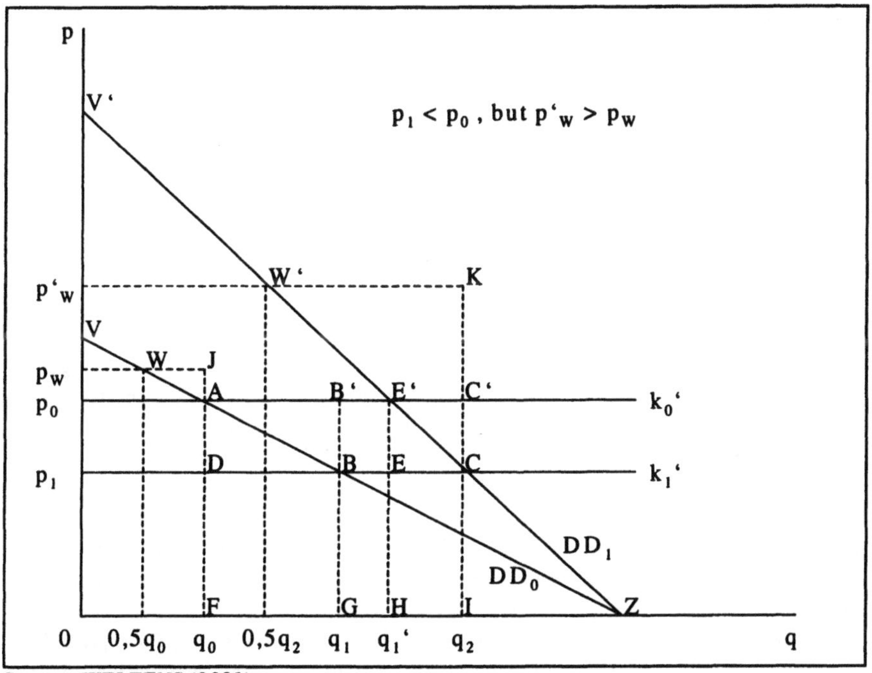

Source: WELFENS (2001)

What is an adequate way to compare the initial and the new value-added? The nominal figures are obvious and one might apply either p_0 or p_1 to the old and new quantity if one is calculating in real terms in a traditional way of deflating nominal figures. However taking into account average marginal utility of users leads to a different price vector for nominal value-added.: p_w initially, thereafter p_w'; here the "value-price" of computers has increased. GORDON's argument that falling computer prices always indicate a falling marginal product of computers must therefore be refuted.

Two-Sector Perspective

Overall output Y is the sum of sectoral outputs, say of ICT output Z and non-ICT output N. Denoting the relative price of non-ICT relative to ICT output as λ we have overall output in terms of ICT output (say consumer goods) is given by:

(A1) $Y = Z + \lambda N$

Therefore, the overall growth rate of output is – denoting Z/Y as v – given by

(A2) $g_Y = v\, g_Z + (1\text{-}v)(g_\lambda + g_N)$

If we have the growth rate of ICT relative to output we can calculate the right hand-side of equation (A2): eg in the US the share of ICT in overall output was 6.1% in 1997 and the annual change of the ICT-GDP ratio was 12.9% (Tab. B2). The product of both figures gives 0.061 * 0.129 = 0.0079; roughly 0.8%. Almost one percentage point of economic growth is related to this sector only. If there are, however, positive technology spillover effects on the other sector the above approach is, of course, underestimating the significance of ICT.

If Z is produced with capital K and labor L_1, according to $K^{\beta}L_1{}^{1\text{-}\beta}$ while $N=nL_2$ – with $L= L_1+L_2$ - we have

(A3) $g_Y = v\, \beta\, g_K + v(1\text{-}\beta)g_{L1} + (1\text{-}z)(g_\lambda + g_{L2})$

An interesting aspect here concerns the rising relative price of non-ICT goods which is raising real output Y (in terms of ICT goods); the fall in the relative price of ICT goods in the second half in the US in the 1990s was close to 25%; the fall in absolute terms was 12% between 1987 and 1994, but 26% in the period 1995-99 (EUROPEAN COMMISSION, 2000, p. 53). Spending on ICT in the US (and Sweden) was much higher than in Euroland countries in 2000 (Tab. B3). From a theoretical perspective spending on ICT imports should contribute to growth if such equipment is associated with learning effects and spillover effects. Hence not only ICT production is important with respect to growth.

As regards differences in expenditures on ICT – relative to GDP – and value-added in ICT, the US difference between expenditures and production clearly suggest that the US is a major net importer of ICT. At the other extreme there is Ireland whose production is much higher than its expenditure on ICT. Both the US and Ireland thus could be suitable countries to clarify the GORDON issue.

As regards opportunities to earn high profit rates in the internet there is limited reason to be very optimistic (and the real winner might be the Old Economy whose firms reorganize the value-added chain more efficiently via the internet):

- markets for information are very imperfect which makes selling information via the net difficult. Creation of specialized user clubs could be a solution to the problem, building reputation on the supply side also;
- barriers to entry are low and actually falling if one takes into account falling chips and computer prices: this can lead to quickly falling profit rates;
- there are considerable legal uncertainties for transactions in the web and there is no truly global market place due to legal problems with international sales; e.g. a German online bookstore will sell a book only to German residents if the internet user wants to pay after receiving an invoice;
- knowledge management systems for the internet are still in the infancy;
- the internet is not really user-friendly as long as information cannot be displayed on a split screen while allowing the user to copy content – with the source as a footnote – directly into word processing and other software applications.

Tab. B2: ICT Sectors in Europe and the US, Value Added (% of GDP)

	1992	1995	1996	1997	1998	1999	95/99 (% p.a.)
B		3,3	3,5	3,5	3,8	4,1	8,6
D	3,5	3,4	3,3	3,6	3,7	3,9	5,1
E		2,8	3,0	3,2	3,4	3,6	12,6
F	3,9	3,8	3,9	4,0	4,1	4,3	6,2
IRL		6,5	6,7	7,5	7,3	7,6	17,8
I	3,4	3,3	3,3	3,3	3,5	3,7	10,3
NL		4,3	4,4	4,5	4,7	5,0	7,9
A		4,7	4,4	4,2	4,4	4,8	1,0
P		3,4	3,5	3,7	4,0	4,3	12,5
FIN		4,3	4,6	5,5	5,5	5,8	21,4
S		4,3	4,8	5,4	5,9	6,5	16,3
UK	5,0	5,2	5,2	5,2	5,4	5,6	7,3
EU11	3,8	3,6	3,7	3,8	4,0	4,2	7,7
US	5,0	5,3	5,5	6,1	6,4	6,8	12,9

Source: CSFB

Tab. B3: Spending on Information and Communication Technology (% of GDP)

	1992	1993	1994	1995	1996	1997	1998	1999	2000
B	5,4	5,5	5,4	5,4	5,7	6,2	6,1	6,7	7,1
Dk	6,1	6,4	6,1	6,2	6,4	6,7	6,6	7	7,3
D	5,3	5,5	5,2	5,1	5,2	5,6	6,2	6,6	7
EL	2,4	2,3	3,7	3,8	3,9	4,1	4,8	5,5	5,9
E	3,8	3,8	3,7	3,7	3,9	4,1	4,5	5	5,3
F	5,7	6	5,7	5,8	6	6,5	7	7,6	8
IRL	5,4	5,2	5,8	5,7	5,8	5,5	5,3	5,3	5,2
I	3,6	3,8	4,1	4,1	4,1	4,3	4,9	5,5	5,9
NL	6,4	6,4	6,3	6,3	6,6	7	7,3	7,8	8,1
A	4,9	5,1	4,5	4,6	4,7	5,2	5,7	6,2	6,5
P	2,7	2,9	4,4	4,8	4,9	5,1	5,8	6,1	8,3
FIN	4,6	5,1	5,5	5,5	5,9	6	6,3	6,8	7
S	7,4	8,3	7,6	7,5	7,4	8,1	9	9,6	10,1
UK	7,1	7,4	7,2	7,8	7,8	7,8	7,5	8	8,4
EUR-12	4,9	5,1	5	5,1	5,2	5,5	6	6,5	6,9
EU-15	5,3	5,5	5,4	5,5	5,6	6	6,4	6,9	7,2
US	7,1	7,3	7,4	7,5	7,7	7,7	8	8,1	8,3

Source: EITO OBSERVATORY (2000); D = Germany

5.2.1 Spillover Effects from Computers Used in Households

SOLOW´s remark that computers are everywhere, but not in the statistics on productivity growth is well known. One neglected aspect of the computer stock-growth accounting puzzle might be related to the fact that a considerable part of personal computers is used by private households where it partly is used for consumption purposes (eg downloading games, sending e-mails). However, computer literacy is a general purpose technology whose diffusion into the household sector should have positive spillover effects on the economy.

Therefore looking at a production function in which only the computers used in firms are considered is likely to underestimate the impact of the computer. The following table shows the computer density of various countries (and the internet host density which is strongly correlated with PC density). There is a strong lead of the US vis-à-vis Japan and the EU. Germany's computer density was about 2/3 of the US figure in 1998. PC density is highly correlated with the use of the internet. The supply side of the digital economy is covered by the internet host density – here Germany is weak by international comparison. The host rate in 1999 was 1/3 of the Dutch figure and an 1/8 of the US figure.

Tab. B4: The Information Age

	Personal Computers per 1.000 people 1998	Internet Hosts per 10.000 people July 1999
Austria	233,4	252,01
Belgium	286,0	266,,90
Denmark	377,4	540,30
Finland	349,2	1.116,78
France	207,8	110,64
Germany	304,7	173,96
Greece	51,9	59,57
Ireland	271,7	156,68
Italy	173,4	68,28
Netherlands	317,6	403,49
Portugal	81,3	59,40
Spain	144,8	76,75
Sweden	361,4	581,47
United Kingdom	263,0	270,60
Japan	237,2	163,75
USA	458,6	1.508,77

Source: WORLD BANK (2000), World Development Indicators, p. 300 and following (www.worldbank.org)

Role of Trade for Growth in the Digital Society

The neoclassical growth model with capital K and labor L as factor inputs can be modified for the case of an open economy in a very straightforward manner where we implicitly take into account recent empirical findings that trade is significantly positively influenced by the intensity of international telecommunications (WELFENS/JUNGMITTAG, 2001) and by the use of the internet as proxied by host density (FREUND/WEINHOLD, 2000). Hence, the expansion of the digital society fosters trade. With a focus on an open economy with exports X and imports X* (* denoting foreign variables) we modify the standard Cobb-Douglas production function – assuming exogenous Harrod-neutral technological progress (dA/dt)/A=a and a constant depreciation rate h - as follows:

$$(I) \qquad Y = K^{\beta}(AL)^{1-\beta}(1+[(X/Y)(X^*/Y^*)])^{\phi}; \ \phi>0$$

The assumption here is that output Y will increase along with rising imports X* normalized to foreign output Y* (alternatively X*/Y could be used). This effect represents the transfer of embodied technological progress via imported goods. Similarly, we assume the export-GDP ratio has a positive output effect since one may expect that rising (normalized) exports facilitate the exploitation of economies of scale and at the same time brings gains through specialization. We get a more compact expression if we assume the standard export equation and import equation to be as follows, respectively:

(II) $X = xY^*$

(II') $X^* = x^*Y$

We can rewrite the production function as follows where we assume x and x* to be exogenous parameters for simplicity:

(I') $Y = K^{ß}(AL)^{1-ß}(1+xx^*)^{\phi}$;

For empirical analysis one might want to replace xx* by $\Sigma(xx^*)$ so that lagged export-GDP and import-GDP ratios are also taken into account as relevant for international technology transfer and specialization gains.

Denoting output per efficiency unit of labor (AL) as y´ and capital per efficiency unit of labor as k' – that is k'=K/(AL) - we have

(III) $y' = k'^{ß} (1+xx^*)^{\phi}$

The growth rate of income per capita (Y/L=y) now is – with g for growth rate - approximately (assuming ϕ, x, x* to be small) given by

(III') $g_y = ßg_k + a + \phi d(xx^*)/dt$

Per capita output growth thus depends not only on the growth rate of capital intensity and the growth rate of technological progress but also on trade intensity - partly itself reflecting technology transfer effects and thus representing an endogenous growth approach - as proxied by xx* (the proposed specification contains a bias, of course, in favor of small open economies which typically have rather high export-GDP ratios). The approach proposed is convenient and easy to handle for empirical modeling. One also may note that x and x*, respectively, could be interpreted in alternative ways, eg if one would assume that the foreign internet user density x* (in country II) has a positive effect on output in country I, while domestic internet user density (x) has a direct positive effect on output.

The crucial differential equation for k' which is capital per unit of efficiency labor is known from standard growth theory (JONES, 1998) to be (with δ representing the capital depreciation rate, s and n being the savings rate and the growth rate of the population respectively):

(IV) $dk'/dt = sy' - (\delta+a+n)k'$

Taking into account the modified production function we therefore have:

(IV') $dk'/dt = s\, k'^{ß} (1+xx^*)^{\phi} - (\delta+a+n)k'$

This is a Bernoulli differential equation which can be shown to be convergent if the parameter with $k^{ß}$ is positive and it will converge to the steady state value given subsequently in (VI). Instead of solving the Bernoulli equation directly we can proceed as follows: The steady state per capita output y# can be obtained from the steady state condition dk'/dt=0 as:

(V) $s\,y' = (\delta+a+n)\,k'$

(V') $sk'^{\beta}\,(1+xx^*)^{\phi} = (\delta + a + n)\,k'$

(V'') $k'^{\#} = [s(1+xx^*)^{\phi} / (\delta + a + n)]^{1/(1-\beta)}$

Using the above production function results in the new equation for the steady state (equilibrium) ratio of capital per efficiency unit of labor:

(VI) $y^{\#} = A_0e^{at}\,\{\{s/[\delta+a+n]\}(1+xx^*)^{\phi}\}^{\beta/(1-\beta)}$

From this perspective we can see that per capita output y is not only positively dependent on the growth rate of technological progress, but also on trade intensity - a refinement of growth theory. We find as a standard result that the depreciation rate and the growth rate of the population reduce the steady state value of y' and y, respectively.

The internet might raise permanently the rate of technological progress and also stimulate trade intensity positively – not least because more services become tradable implying an increase in import competition. According to MANN (1998) rising import competition stimulated productivity growth in the US in the 1980s; by contrast there was – for unclear reasons (possibly due to labor market rigidity) - no significant productivity impact from the rising import-GDP ratio in Germany. Note, to the extent that the internet reduces trading costs the average market radius will increase which lets us expect a rise in firm size and a parallel increase of exports.

ICT Production and ICT Output Effect

How can one easily take into account the use of ICT as opposed to the output effect of producing ICT? A very simple way to do this is to assume the following production function in which Q, the stock of ICT capital used in firms plays a role, other real capital K, labor L and the stock of imported computers ΣJ – with J standing for computers imported per period; for simplicity we assume that J is an infinitively lived equipment used by households which learn computer skills by using such equipment at home. To the extent that those households are not retired there should be spillover effects as people learning at their home computers – and using the internet – will contribute to aggregate output and factor productivity, respectively. This spillover effect for the firms basically is captured by the factor [Σln J]/L in the following production function.

(1) $Y = Q^{\alpha}K^{\beta}L^{\gamma}[1+((\Sigma \ln J)/L)]^{\varphi}$

Using continuous time and assuming that at the start of the economy, in t=0 there was a "natural import" of unity we can calculate the integral for lnJ = J(ln J - 1) as being equal to J_t [(lnJ_t −1) – 1/J_t]. Taking logarithms – and assuming that (Σ lnJ)/L is rather small so that we can use the approximation ln (1+x) ≅ x we can rewrite the above equation as the supply-side of the economy:

$$\text{(2)} \qquad \ln Y = \alpha \ln Q + \beta \ln K + \gamma \ln L + \varphi\{[\ln J + \ln(\ln J - 1 + J^{-1})/L]\}$$

If $\gamma = 1 - \alpha - \beta$ we can rewrite (2) as follows (using y=Y/L; q=Q/L; k=K/L):

$$\text{(2´)} \qquad \ln y == \alpha \ln q + \beta \ln k + \varphi\{[\ln J + \ln(\ln J - 1 + J^{-1})]/L\}$$

From this it follows that per capita output should depend on the growth rate of domestically produced computers per capita (computers used in firms), the growth rate of other capital per capita and a rather complex term for imported computers per capita.

The equations (2) and (2´) can be estimated empirically. We expect a positive sign for the impact of imported computers. The potential misspecification in GORDON´s approach mainly refers to the term { } in equation (2´). For empirical purposes one would, of course, look into various lag patterns for both computers used in firms and in households.

5.2.2 Endogenous Growth in an Open Economy

One option to model growth in an open economy is a neoclassical approach in which trade intensity affects only the level of the growth path. One can also, however, consider a direct effect on the growth rate. In an open economy with export and import of goods and (digital) services one may model the impact of trade in the following way: We use the basic neoclassical growth model in a setting with zero population growth (n=0), and assume that exports equal imports so that the investment-GDP ratio is equal to the savings rate. The growth rate of labor-saving technological progress (a=dA/dt/A) can be modeled – with R/L as the exogenous ratio of researchers R to the total labor force L - as

$$\text{(A.I)} \qquad da/dt = \{[(X/Y)(X^*/Y^*)]^{\theta} (R/L)^{\omega}\} a^{\Omega} - \mu a$$

Exponential parameters are assumed to be positive, and $0<\Omega<1$. The equation shows that we have assumed that technological progress declines over time in accordance with a depreciation rate μ. However, based on the existing growth rate of technological progress there could be an increase of a if certain amplifiers are present: We assume that labor saving technological progress is positively related to the export-GDP ratio which could reflect the impact of economies of scale in labor-intensive industries; moreover, the foreign export-GDP ratio, that is the ratio of imports J to Y* (domestic imports J are equal to foreign exports X*=jY) also has a positive impact on the growth rate of technological progress - reflecting international transfer/import of labor-saving technological progress and of labor-intensively produced goods. While in reality one might argue that X*/Y* are not the major factors of importance, rather the intensity of import competition as proxied by J/Y, we have chosen here a specification which reflects the basic idea that imports matter for progress. But our formulation allows for relatively easy mathematical manipulation. Assuming standard export and import functions X=xY*, X*=x*Y we can rewrite the above technological progress function as

(II.A) $da/dt = \{[xx^*]^{\theta} (R/L)^{\omega}\}a^{\Omega} - \mu a$

The solution of this equation is

(III.A) $a(t) = \{C_o e^{-\mu(1-\Omega)t} + [[xx^*]^{\theta} (R/L)^{\omega}]/\mu\}^{1/(1-\Omega)}$

For this equation there is - assuming $0<\Omega<1$ - a steady state solution a# which is:

(IV.A) $a\# =\{[[xx^*]^{\theta} (R/L)^{\omega}]/\mu\}^{1/(1-\Omega)}$

Taking into account the production function $Y = K^{\beta}(AL)^{(1-\beta)}$ we get a steady state for Y/L, namely:

(V.A) $y\# = e^{at} [s/(a+\delta)]^{(\beta/(1-\beta))}$

(VI.A) $\ln y\# = t\{[[xx^*]^{\theta} (R/L)^{\omega}]/\mu\}^{1/(1-\Omega)} + [s/(a+\delta)] (\beta/(1-\beta))$

(VII.A) $y\#=e^{t\{[[xx^*]POW\theta \; (R/L)POW\omega] /\mu\}\}POW(1/(1-\Omega))} [s/(a+\delta)]^{(\beta/(1-\beta))}$

Thus the export ratio and the import ratio as well as R/L – reflecting relative research input – directly and positively affect the trend rate of technological progress and hence the long term growth rate of y. The savings rate s and the depreciation rate on capital will affect the level of the expansion path of per capita income.

The link to the internet is straightforward as one may assume that the internet makes more goods and services tradable which should not only facilitate international technology spillovers but also the exploitation of static and dynamic economies of scale in the tradables sector. Hence the switch to the digital internet society can be understood as one in which x and x* are increasing, possibly also the parameters θ and ω such that there would be a long term increase in the growth rate. Note that in principle x and x* might also be interpreted in a different manner directly related to the ICT sector, eg. as internet host density at home and abroad, respectively. In conclusion, one might consider x and x* to measure the internet user density at home and abroad. International network effects provide a good ground in this context for using the specification xx* - rather than x+x* .

Thus we have a straightforward contribution to endogenous growth theory. The basic hypothesis is that the use of the internet and new digital equipment (eg digital switches and faster PCs) is permanently increasing long term growth. As regards innovation and diffusion, one may anticipate that a pro-entrepreneurial politico-economic framework is helpful in creating higher economic growth. This is due to the tendency of new firms to adopt modern, faster technology than incumbent operators. With respect to the diffusion of US digital telecommunications this was shown by SHAMPINE (2001).

There is a caveat since empirical research would also have to clarify whether the depreciation rate μ is raised by the internet as the diffusion rate of new knowledge is accelerating in the digital age and the New Economy. If the long term growth rate would increase by just ½ percentage point due to the growth of the internet and new digital services – largely linked to modern telecommunications and the use of computers – this would certainly have a large impact on the OECD countries and respectively the world economy. It is unclear whether economic cycles in a world economy with higher economic growth will be more accentuated than in the past, hence we face new challenges for research which are not least linked to the issue of how the internet will affect price adjustment speeds in goods (and labor) and assets markets.

One may note that a model with two sectors can be useful for an analysis; incidentally it can be shown that an explicit two-sector model suggests that the standard Solow model overestimates the role of the savings rate for growth (WELFENS, 2001 b). If one is analyzing the new economy without taking into account the fact that computer installed – namely computer produced plus net import of computers (read here capital) – is the relevant variable, one would have a pitfall in modeling; GORDON (1999) has indeed made this pitfall in his study on the new economy in the US. In a paper of COLECCHIA/SCHREYER (2001) there is evidence that both in the US and in some other OECD countries there are positive effects of ICT capital investment on economic growth. ICT diffusion plays an important role, too - not only is the ICT production sector relevant but also the policy framework.

5.2.3 Network Effects and the Macroeconomic Production Function

As is well known the use of computers and modern telecommunication equipment often is characterized by network effects. For the initial users the network becomes more useful with additional users linked to it. Therefore users might actively promote the use of their network, but more importantly, networks – say AOL, MSN or YAHOO (giving some incentive to existing users) – have an incentive to encourage users to convince friends and family to also use the network. Network effects also can play a role in industry, especially if there are strategic technology alliances in technology intensive industries. Indeed, the number of technology alliances particularly strongly in the US in the 1990s (COUNCIL OF ECONOMIC ADVISORS, 2001, p. 117).

A logistical function can be used to model network effects in the new economy, that is we can use the standard innovation diffusion model. Assume that Z(0)=1, the adoption rate for the innovation Z is given by the differential equation:

$$\text{(C.1)}\quad dZ/dt = bZ(L-Z)$$

L is the exogenous population, Z the number of persons using the innovation - say a mobile phone, a PC or the internet - and L-Z the number not using it. Whenever pioneer users meet nonusers there is an "infection" effect which is described by the positive parameter a. In an economy which also is a heavy producer of the innovative product the diffusion parameter a is likely to be higher than in a setting where the innovative product is only used.

The solution of the above equation is given (see BECKMANN/KÜNZI, 1982, p. 130) by a logistical equation for the stock variable Z(t):

$$\text{(C.2)}\quad Z(t) = L/[1 + (L-1)e^{-bt}]$$

Note here that in an open economy the parameter a might be influenced by both the export-GDP ratio (x) and the import-GDP ratio (j) to the extent that the "infection" rate in the tradables sector is more intensive than in the nontradables sector; a suitable function could be $b= b_0(1+ x+j)$. If the presence of foreign direct investment (FDI) at home and abroad reinforces international diffusion of knowledge an FDI stock proxy K#/Y and K*#/Y* for the stock of FDI at home and abroad relative to GDP might also affect the diffusion rate. Moreover, one could study an export function with standard products (1...n) and innovation products (m...z) where the innovation products are ranked according to novelty – with novel products assumed to fetch higher prices than standard products; innovation products are increasingly adopted by firms in the export sector. After some critical time T innovation products become standard products with no premium in the world market, that is prices are equal to marginal costs. The terms of trade thus will depend on the percentage of innovation products in the overall basket of export goods. We will not look in these special problems here. Rather we will continue with the above equation C2. Denoting Z/L as z we have

$$\text{(C.2´)}\quad z(t) = 1/[1 + (L-1)e^{-bt}]$$

Assume that there is a dual use good (dual here refers to its double nature as consumption and investment good) – here mobile telecommunication devices that are first bought as consumption goods – whose diffusion is described by (C.1) and (C.2), respectively; we use the following modified production function in which positive spillover effects from mobile telecommunication devices Z are entering:

$$\text{(C.3)}\quad Y = K^{\beta}L^{1-\beta}Z^{\sigma}$$

The basic assumption here is that the use of computers in households has positive spillover effects in industry; there are knowledge spillover effects into output of firms whose aggregate value-added is Y. A special case which is easy to handle is where $\sigma=1$:

$$\text{(C.4)} \quad Y = K^{\beta}L^{1-\beta} L/[1 + (L-1)e^{-bt}]$$

In per capita terms we have (with y=Y/L and k=K/L):

$$\text{(C.4')} \quad y = k^{\beta} Lz(t)$$

Interestingly, the size of the country plays a positive role for per capita income here. From this perspective there could be an advantage for large countries – with large markets (US or EU or other regional integration areas) – in the era of the New Economy. In such a model setup the size of the country matters. Comparing the NAFTA and Europe there is no reason to believe that Europe cannot benefit from size effects as much as the North America plus Mexico.

An important aspect concerns the impact of the ICT boom on the wage-real interest rate-ratio. If ICT is partly Solow-neutral capital augmenting technological progress the ratio of wages to interest rates will increase which will stimulate US firms to relocate ICT production abroad. Say to Mexico, Ireland or Hungary or Malaysia. This means that ICT production in low income countries will grow so that the growth-enhancing effects of strong productivity growth in ICT finally arrives outside the pioneer country (US, partly Japan). If the host countries for foreign direct investment in ICT should reflect as small a group of host countries as for FDI in general in the 1980s and 1990s only two dozen countries worldwide will benefit from ICT product cycle trade.

In a Schumpeterian economy there will be a series of innovations coming to the market. If such innovations are not evenly distributed over time they can cause a spurt in growth when a cluster of innovations is entering the diffusion stage, later the growth rate will slow down. If this stimulates entrepreneurs to come up with new innovations – again being realized in a cluster-type fashion at some point t – there will be another economic upswing. Instead of using a single differential equation for one innovation, one will have to use a system of logistical equations for all products i (i=1...n). The new economy is generating logistical diffusion patterns both in the household sector – namely in the case of mobile telecommunication equipment and PCs plus internet use – and in industry (computers, internet use).

5.2.4 Impact of the Internet in a Model with Human Capital and Unemployment

In the following growth model with human capital H we largely follow JONES (1998, p. 48-50) as we use a production function, but we will deviate from his approach by taking into account unemployment and taxes.

$$\text{(D.1)} \quad Y = K^{\beta} (AH)^{1-\beta}$$

Where A is labor-augmenting technological progress growing at an exogenous rate a.

Assume that unskilled labor L is learning skills for time z (exogenous variable) generates skilled labor H according to

(D.2) $H = e^{\varepsilon z} L$

where ε is a parameter and e the Euler number; the time z required for learning might fall in the internet society if households use the internet mainly for additional leisure purposes, alternatively the main effect could be an increase in the efficiency parameter ε: the internet makes learning more effective.

We will assume that those who are unemployed are not saving and therefore the aggregate per capita savings rate is s(1-u) where u is the unemployment rate. Taking into account the role of the unemployment rate and of the tax rate is the only modification which we introduce here as compared to JONES (1998). We thus replace the normal savings rate s´ by $s(1-u)(1-\tau)$; the assumption here is that those unemployed have zero savings, that is s actually is the savings rate of workers. Moreover, workers save out of income net of taxes. Physical capital is accumulated according to

(D.3) $dK/dt = s(1-u)(1-\tau)Y - \delta K.$

We use lower case letters to denote variables divided by the stock of unskilled workers L and thus can restate the production function in terms of output per (raw) worker as

(D.4) $y = k^{\beta}(Ah)^{1-\beta}$

where h is defined as $h= e^{\varepsilon z}$. Let us define state variables by dividing by Ah and denote this variable with ´#. The steady state output variable is

(D.5) $y'\# = k'\#^{\beta}$

Capital accumulation in terms of the state variable is given by

(D.6) $dk'\#/dt = s(1-u)(1-\tau)y'\# - (n+ a + \delta)k'\#$

The steady state values of k´# and y´# are found by setting dk´#/dt = 0 which results in

(D.7) $k'\# / y'\# = s(1-u)(1-\tau)/(n+ a + \delta)$

Substitute this result into (D.5) to obtain that the steady state value is

(D.8) $y'\#\# = [s(1-u)(1-\tau)/(n+ a + \delta)]^{\beta/(1-\beta)}$

Output per worker employed therefore is in the steady state

(D.9) $y(t) = [s(1-u))(1-\tau)/(n+a+\delta)]^{\beta/(1-\beta)} h\, A(t)$

We define per worker income relative to the US as

(D.10) $y'' = y'\# / y'\#_{US}$

and obtain that relative income y" is given in a world with identical production functions as

(D.11) $y'' = (s''E''T''/Z'')^{\beta/(1-\beta)} h''A''$

where $E''=(1-u)/(1-u^*)$, $T''=(1-\tau)/(1-\tau^*)$, $Z=(n+a+\delta)$ and all relative variables carry ". The main difference compared to JONES (1998) is the impact of equilibrium unemployment and of the income tax rate.

Notice that relative income will not be constant unless we have identical progress rates a=a* where * stands for foreign country (here the US). Long term differences in international per capita income – the transatlantic per capita differential in the 1990s – positions reflect in this perspective

- low savings rates outside the US; here this means Europe, and indeed the savings rate in the EU has fallen considerably in the 1990s; part of this fall actually could be related to the Euro since households and firms will face lower requirements in terms of equity capital when borrowing in a more competitive financial market;
- a higher unemployment rate; the unemployment rate in Europe is higher than in the US (at least since the 1980s) and the high flexibility of the US labor markets has helped to translate the ICT potential for expansion into a lower unemployment rate in the US – much in contrast to the EU core countries Germany, France and Italy;
- a higher income tax rate; here tax reforms in EU countries have helped Europe to catch up with the US to some extent, however, it is unclear whether EU countries are well prepared to cope with international pressure for even lower income tax rates – here the internet plays a role since it will make taxation of income and value-added more difficult than taxation in the 20th century.
- a higher rate of population growth; with respect to comparing the EU and the US, the problem of a higher rate of population growth actually is on the side of the US;
- a higher depreciation rate;
- a lower investment effort in human capital as indicated by h/h_{US}. One may note that h and h* are increased by the use of the internet to the extent that diffusion of knowledge and accumulation of human capital is facilitated by the internet. The US leadership in the internet thus should translate into a

relative rise of the US per capita income. Moreover, the US also has established a clear lead in growth of expenditures on education in the 1990s;
- a lower endowment with the initial level of technology as indicated by A and A", respectively.

5.3 Conclusions and Policy Implications

The GORDON view of computers and ICT is doubtful for several reasons. But further research is needed. From the human capital growth model it follows: Europe will fall behind the US in terms of per worker income if EU countries are not reducing unemployment rates and reducing tax rates and raising efforts on human capital formation. The most difficult problem in Europe is to finance higher education in private universities. European governments will find it very difficult to generate higher tax revenues in the era of globalization and the internet –; unfortunately the idea of private universities (and vouchers) is unpopular in Europe. The competition of private universities would not only mean a higher stock of educational capital but also would stimulate the efficiency of teaching and research efforts in public universities. An economically meaningful investment-output ratio in the knowledge society should be defined as the weighted sum of the traditional investment-output ratio, the R&D-GDP ratio, the software expenditure-GDP ratio and the education expenditure-GDP ratio. In the latter three fields the US has a clear lead over Euroland and Germany, respectively. Germany might benefit from the large size of the home economy and from high ICT investment of big firms from the Old Economy. It is, however, doubtful that there will be relatively many German e-companies which could become leading international players in the future. Government has not made expansion of the use of the internet a policy priority, indeed the federal government has been unwilling to spend money on bringing the schools on the internet, rather government-owned Deutsche Telekom was invited to sponsor internet access.

How about policy priorities in the EU? Taking a look at the main budget items of the European Community – with No. 1 agriculture (share about 45%) and No. 2 structural funds (share about 1/3) – we do not find adequate priorities. Agricultural subsidies should be reduced to less than 20% of the EU budget while structural funds should be reformed in a way that would support more strongly retraining, education and support for research and development.

If both the production of ICT and the use of ICT are important for economic growth there could be two ways for high growth. Trying to become a country that has successfully specialized in ICT production or encouraging firms and households to quickly use the new technologies. As regards the use of new ICT and promotion of the diffusion process governments could have a role in countries which are not leaders in ICT production – namely to the extent that government can substitute for the diffusion impulses which come from ICT production in other countries. The Scandinavian countries, the Netherlands and Germany have embraced telecommunication liberalization energetically after 1998, the UK already before. This lets one expect that part of the EU could be a driving force in the

European and global ICT revolution. However, while the EU partly looks strong in the field of telecommunications its role in the fields of computer and software looks relatively modest when compared to the US.

It would be useful to study the user patterns of internet users more closely – both with respect to the time budget and the structure of use. Moreover, the size of network effects and the significance of the role of spillover effects from household computer use should be estimated empirically. Part of the problems encountered here will refer to data problems as neither government nor industry has straightforward data which could tell where computers sold finally are used. But the data issue should be solved by survey analysis and other methods.

We have shown some basic ways to analyze the New Economy in a meaningful way. This analysis is, however, only a modest beginning which invites to more advanced modeling and broad empirical analysis.

5.4 High Dynamics of Telecommunications and Productivity Growth

The field of modern telecommunications and of the internet plus multimedia is characterized by high technological dynamics. This brings decisive advantages for innovative flexible firms, and this often means opportunities for newcomers and small firms. Established big firms mostly are not flexible enough to accommodate the high speed of innovations in the internet and multimedia sector.

Tab. B5: Firm's Use of Modern Media in the late 1990s

	CD-ROM for information	Internet for information	CD-ROM for marketing	Internet for marketing
USA	80	93	23	>50*
UK	75	91	23	47
Germany	72	83	19	45
Japan	69	93	6	27
France	58	83	16	25

Source: Information Society Initiative (1998) and author's estimate..

Both US and British firms, with respect to acquiring information and to marketing, use new information technologies more often than German firms – with the exception of the use of the internet in marketing. This reveals the extent of opportunities which exist in the better usage of the CD and internet among continental EU countries.

English being the main language on the internet, this does not rule out that in most cases homepages in both national and English exist. In the US, 3 out of 10 firms have internet shopping activities; in Germany, only 16% of firms engage into such activities; in the UK and Japan, the respective figure reaches 13%. An adequate design of the homepage is crucial for the firm as is the development of a clear profile of its internet activities. Communication via the internet is typically by nature rather short and the user can therefore be guided by key words. Long texts are unpopular and in such cases, downloading options often exist. While graphical elements are a quality element of the website, practical considerations still constitute a caveat: as long as there is no broadband internet available – in principle this could be quickly established on the basis of cable TV – the display of graphics on screen will remain a time-consuming task. An analysis of the extent to which different sectors have make use of the internet and their perceptions of the tool remains to be done. In principle the internet can be very useful for the firm in several fields as it is shown in the following graphics. Telecommunication-based services could in the future play an enormous role among OECD countries. Expected sale figures for e-commerce show a rapid growth in the US and the EU for the period 1995 – 2005.

Telecommunication Productivity Growth

An important aspect of modern information and communication investment is that investment in computers and new software should raise telecommunication productivity growth considerably. Conventional output and productivity measurement underestimates the growth effects of computers and telecommunications to some extent (SCHREYER, 1998). There is a problem with the measurement of real output in information, computer and telecommunication-producing industries and the share of these industries in the overall output is increasing over time. The share of the G-7 countries in manufacturing value-added increased from 8.7% in 1980 to 11.6% in 1995. Quality adjustments of final expenditure measures show that the share of expenditure on information and communication products has not stagnated over time but increased in selected OECD countries. The increasing use of information and communication technologies in work encourages labor productivity growth considerably (OECD, 1998). It can be further noted that the use of internet does stimulate the dissemination of knowledge.

Estimating Productivity Growth at the Macroeconomic Level

Technological progress is crucial for growth and the dissemination of knowledge and the accumulation of new knowledge is decisive for prosperity. It is also undisputed that information and communication are becoming increasingly important as comprehensive functions in the economic order. Telecommunications play a central role in this. At the beginning of the 90s, the global market volume of the information services amounted to approx. 500 billion ECU, whereby 330 billion ECU went to telecommunication services and approx. 170 billion ECU to software and computer utilization (cf. WELFENS/GRAACK (1996), p. 9). These figures are expected to increase sharply in the long-term. In clear contradiction with the theory and economic relevance, however, is the question regarding the level of technology responsible for the technological progress and the role of information and communication in macro-econometric production models. So when estimating production functions (e.g. a Cobb-Douglas production function), technological progress is commonly approximated by a linear time trend. This procedure reveals a series of weaknesses. On the one hand, the inclusion of a time trend does not provide an explanation for technical changes, i. e. the causes or sources underlying technical progress are not distinguishable. At the most, the order of magnitude of the technical progress can be estimated. On the other hand, the changes in the rate of technical progress are not identifiable and technical progress is therefore presented as an unexplained uniform growth. Only a few authors have attempted to take pay greater attention to technical progress by using more appropriate indicator variables (cf. BUDD/HOBBIS, 1989; BUDD/HOBBIS, 1989a and COE/MOGHADAM, 1993). Other authors (cf. SCHRÖER/STAHLECKER, 1996) approximate changes in technical progress by using dummy variables which change the slope of the time trend. A formal record of the influence of information and communication in macro-econometric production functions by means of appropriate indicator variables is almost inexistent. To the best of our knowledge, only RÖLLER/WAVERMAN (1996 and 1996a) have investigated empirically within a production function framework how telecommunication infrastructure affects economic growth.

Within large industrial enterprises and among smaller firms, telecommunication has greatly facilitated the production process especially within firms dealing with internal networks such as multinational enterprises dealing with projects in different geographical locations and time-zones. Until 1998 when telecommunication networks in Europe were opened to competition, EU-firms were entirely dependent on public networks due to the existence of monopolies - state-owned or partly privatized telecommunication enterprises. Information inputs are particularly important for the supply of the service sector, especially for banks (cf. WIELAND, 1995). Major international banks, for which information and communication is a direct production input, may spend up to 30 % of total costs for telecommunications.

The transformation process of the Eastern European countries shows us that a capable telecommunication sector is essential for the building of a market economy responsible for achieving efficiency and productivity gains. In a market economy, information about changed input and prices, technologies and market expectations are exchanged within seconds via telecommunication networks. Furthermore, the optimal use of knowledge will become more important in an increasingly knowledge-based society (cf. OECD, 96). Thus, the interest in the elasticity of production of the factor information grows.

During the sixties and seventies West Germany showed a relatively low telecommunication density (number of main connections per 100 inhabitants in 1990: West Germany 50, USA 60 and Sweden 70) and high call charges, mainly due to the monopoly and the less internationalized supply relations of the German "Bundespost". Unfavorable preconditions for optimal information processes within the West German economy prevailed. In comparison with the USA, characterized by high competition since the breaking-up of AT&T in 1982, West Germany was lagging behind in the area of telecommunication technology by the end of the 80s – calculated from the number of relative patent applications - and further showed a negative comparative advantage when studying the exports of telecommunication goods (cf. GRUPP/ MÜNT/SCHMOCH, 1994).

In the WELFENS/JUNGMITTAG study (1998), alternative sources of technical progress were identified and approximated by means of indicator variables, which were then considered when estimating long-term production functions for the business sector of the Federal Republic of Germany, excluding from the calculations agriculture, forestry, and fishing as well as flat/housing rental for the period 1960 -1990. It also was distinguished between technical progress which is the result of private research and development activities, and the import of technological know-how through licensing agreements.

To be able to study properly technical progress (within an empirical analysis), the JUNGMITTAG/ WELFENS (1998) study considered the time factor regarding the stock of patents at the German Patent Office (Deutsches Patentamt). Secondly the study included the real fees for licenses captured in the balance of payments of the Federal Republic of Germany. In addition to these 2 variables, the use of telecommunications in the long-term production function so that it will be approximated thanks to the indicator variables i.e. the number of annual telephone calls.

For estimating the long-term production functions, the concept of the co-integration of time series introduced by Engle and Granger (cf. ENGLE/ GRANGER, 1987) was used by JUNGMITTAG/WELFENS, 1996. For various reasons the three-step procedure, proposed by ENGLE and YOO (cf. ENGLE/ YOO, 1991), was used by JUNGMITTAG/WELFENS, 1998 to remedy shortcomings of conventional estimations. The third step of ENGLE and YOO, added to the Engle-Granger two-step procedure, provides a correction to the parameter estimates of the first stage static regression which makes them asymptotically equivalent to FIML and provides a set of standard errors which allows the valid calculation of standard 't' tests. The superior long-term production function was then used in order to roughly assess the effects of the technical progress approximated by the indicator variables and of the need for information and communication, approximated by the number of telephone calls, as well as the impact of the usual production factors on economic growth from 1961 until 1990 for the case of Germany. Some main results are summarized in the following table.

Tab. B6: Explaining the Growth of Real Gross Value-Added in the Enterprise Sector in Germany

Source of Growth	Average annual growth rate						
	61 - 90	61 - 65	66 - 70	71 - 75	76 - 80	81 – 85	86 – 90
k_t	1,5	2,5	1,9	1,8	1,1	0,8	1,0
l_t	0,2	0,7	0,1	-0,6	0,5	-0,6	1,1
pat_{t-3}	0,1	0,3	0,2	-0,3	0,2	0,1	0,3
la_t	0,3	0,4	0,3	0,3	0,0	0,0	0,8
tv_t	1,1	1,2	1,7	1,2	1,3	0,7	0,7
Total: Estimated	3,3	5,2	4,3	2,3	3,2	0,9	4,0
Actual	3,3	5,2	4,4	1,7	3,6	1,1	3,8

Note: The sum of the individual growth rate components may diverge from total estimated growth rates due to rounding; k = capital stock, l = labor volume, pat = patents, la = real expenditures on licenses, TV = number of telephone calls.

Source: JUNGMITTAG/WELFENS (1996, 1998)

The results show that about 1/5 of economic growth in Germany can be explained by the use of telecommunications – hence a modern and cheap telecommunication sector is absolutely crucial for long-term growth. The results for Germany are statistically robust and suggest similar conclusions for the whole EU.

In contrast with most empirical studies, technical progress in the JUNGMITTAG/WELFENS (1998) study is not approximated by a simple time trend but, the alternative sources of technical progress are taken into account by more appropriate indicator variables: the results of personal research and development from the lagged stock of patents and the import of technological know-how from the real license expenditures. In addition, the importance of information and communication as a comprehensive function in the economic order, showed to a high degree by the increased use of telecommunication services, is integrated in the production function.

After undertaking different test procedures, the empirical results suggest that a long-term production function, which includes all three indicator variables and with the condition that the sum of the partial production elasticities of the factors capital and labor equals one, is the superior specification. It is superior to the simple production function as well as to any augmented production which includes only one or two indicator variables.

This restricted production function is then used to roughly assess the effects of the influences approximated by the indicator variables as well as the impact of the usual production factors on growth of real gross value-added. Here, it becomes clear that the development of the capital stock has the greatest impact on economic growth. But the role of information and communication, approximated by the telephone calls, is also very important.

C. Telecommunications Policy in the European Union

1. Introduction

After a century of a state dominated telecoms sector Europe is facing the challenge of introducing competition in the telecoms sector. The EU has set 1998 as the date for liberalization of voice telephony and network operation - with extended transition periods for Greece, Ireland, Luxembourg, Portugal and Spain. In the UK competition was already introduced in 1984, using a duopoly approach. But in 1995 the former monopoly operator British Telecom (BT) still held about 85% of the British market. Since then newcomers from the cable TV sector (including US companies Nynex, Cablecomms, Bell Cablemedia, which joined forces with Cable & Wireless) have increasingly moved into the telecoms business. This move was possible since digitization has blurred traditional market demarcation lines between video, audio and voice transmission, reinforcing competition in telecommunications. However, BT found it difficult to remain a profitable company in the more competitive environment; it hived off the mobile subsidiary as a separate joint stock company, and the global cooperation with AT&T also fell apart in 2001. This shows that building a global alliance is not a straightforward efficient answer to enlarged international markets. The cooperation between France Telekom, Deutsche Telekom and Sprint also turned out to be unstable.

The telecoms sector is of strategic importance to the whole economy since new technologies offer new communication options and because modern telecommunications could reduce information and transaction costs in the whole economy. Faster and cheaper information will facilitate the creation of more productive company networks across countries. Thus modern telecommunications is not simply reinforcing the role of markets but firms can grow, too, as new forms of internal organization can be realized. They can be organized in a more decentralized way such that the business sector becomes more flexible. By facilitating learning about new technologies and other innovations as well as providing firms with new opportunities for finding optimal commercial partners telecommunications stimulates economic growth as empirical evidence for Germany shows (JUNGMITTAG and WELFENS, 1996). The importance of telecommunications for economic development has been emphasized in many recent studies (e.g. WELLENIUS et al., 1993).

For economic development the provision of infrastructure is crucial which brings positive effects on employment and income during the construction period. Later when infrastructure is used there will on the one hand be positive productivity effects on the user side. On the other hand, a high performance infrastructure will enlarge the market radius and thereby intensify competition. If infrastructure is provided by government for free the infrastructure induced productivity gains for real capital owners implies effectively redistribution effects in favor of real capital owners (RICHTER/SEITZ/WIEGARD, 1996).

To the extent that government provided infrastructure as a public good for firms they will appropriate an economic rent (extra profits which to some extent will be shared with workers). The value of firms with generous access to infrastructure will increase relative to other firms. Traditionally it is assumed that infrastructure is financed via taxes. If government provides infrastructure - e.g. telecommunications - at a monopolistic price the rent from infrastructure accruing to firms is reduced accordingly. The difference between the monopoly price and the price under competition is equivalent to a tax. Switching to a regime with competition implies in the case of telecommunications falling telecoms prices and improved services, which implies windfall profits for firms using telecommunications services. Such windfalls profits particularly benefit firms in communication intensive sectors, e.g. banking and insurance as well as media. In the long run competition will force firms to reduce prices and hence to pass costs advantages to the ultimate customer. It is important to notice that all firms use telecommunication services as an input for production so that competition in telephony will bring down telecoms prices at first, but also the prices of all products in the long run.

Infrastructure which is based on networks - such as streets, railways, electricity and telecommunications - has some special characteristics. Industries based on a fixed physical network are capital intensive and raise the problem of political excess investment which means that lines are doubled although there is no necessity for this. In telecommunications the potential inefficiency of double lines or underutilized hardware under competition has to balanced against the benefits from competition, including a higher rate of innovation, improved services and lower prices (plus thereby induced additional demand). The argument that falling average costs - the case of a natural monopoly - will lead to a deficit problem is not valid since a two-part tariff can be introduced which recovers fixed costs in the form of a monthly user fee (club price) and offers services in accordance with marginal costs. Regulation of network industries in combination with a state owned monopoly was considered in most European countries as the answer to the natural monopoly problem of network industries. Regulations concerned investment, pricing and market access; in telecommunications also a catalogue of universal service obligations and uniform pricing for households and firms. Interestingly, network operation was not organized as a separate firm but as a vertically integrated state monopoly. Whether railways, telecoms or energy, vertical monopolies were allowed. Thereby not only competition in network use was ruled out but also the competition with respect to network services. For the use of railway networks and of telecoms networks in Europe it held until the early 1990s that services were overpriced. Similar observations were made in electricity, where continental European users - in contrast to the British example of competition among electricity generators - could not get additional information „services" about the exact use over time of electricity in individual plants. Such information would be useful to identify least cost technology.

The usefulness of a network very much depends on avoiding bottlenecks so that there are no congestion costs. Such bottlenecks have to be overcome by adequate investment or pricing which differentiates during day and night in a way that mitigates peak demand (peak load pricing). A high performance network will emerge only if individual networks can be combined to a seamless net. Diverging technical standards are a potential problem in this respect - well known in British vs. continental railway systems. If users have a choice among alternative networks there will be pressure for price cuts. The weakest network will indicate the price floor in the competition process (in transportation, taking into account intermodal competition, this usually is the price of railway).

Following the European Commission's initiative alternative telecoms network operators - railways, gas, electricity, water companies and high network operators - will be able to open up their firm internal communication networks to other users. What holds for data transmission since 1996 also holds for voice telephony after 1998 so that EU developments are fare-reaching: deregulation, privatization - and creation of new firms - and internationalization characterize EU telecommunications (WELFENS/GRAACK, 1996).

Modern telecommunications is technology intensive since it relies on computerized exchange, digitized signal transmission and fibre glas lines as well as wireless technologies. As multinational companies (MNCs) are known to be particularly active in technology intensive sectors it is clear that telecommunications is a natural field for investment of MNCs, especially when it comes to modernizing the telecommunications sector in a liberalized regulatory regime. In Europe the process certainly will differ from the development in Latin America in the 1960s when AT&T took over national monopolies and established a new private one - with little benefits in terms of efficiency gains, modernization, price reduction or market growth. EU developments in terrestrial networks are more likely to follow the course of mobile telephony in the European Union which was liberalized in the early 1990s. With several competitors active in mobile telephony in several EU countries services improved and markets expanded while prices fell. Mobile network operators which partly rely on the use of the dominant operator's network face the problem that monopolistic access charges of the dominant operator imply higher prices for mobile telephony than otherwise so that market growth is impaired. This is the interconnection problem of telephony, where the dominant operator certainly has less interest in interconnections than the newcomers with their rather small customer bases. With competition in long distance telephony one may expect as a major benefit falling mobile telephony prices. This will occur in Western Europe mainly after 1998, the liberalization deadline set by the EU; except for the UK and Scandinavia (Sweden, Finland) which already have liberalized their telecoms networks. Since 1984 the US where AT&T was dismembered benefits from competition in long distance telephony so that both the leading three long distance companies and the seven so-called Baby bell companies - with strong regional market positions - are available for international joint ventures and foreign direct investment involving a major equity stake in a telecoms operator.

There cannot be true competition without privatization. Resistance against privatization is likely to be considerable since workers and managers of the incumbent operator fear changes under competition. Moreover, government bureaucrats are losing power when firms are privatized. However, an optimal liberalization will lead to major economy-wide efficiency gains, positive growth and employment effects and higher tax revenues in the long run. Aggregate effects are more important than sectoral effects.

Pricing Problems

For decades the telecoms sectors was a sheltered business in EU countries which was provided by national telecoms monopolies. It was argued that falling average costs in network operation implied a natural monopoly (as a matter of fact, natural monopoly requires subadditivity of costs). Falling average costs imply an extreme case of economies of scale so that the fastest growing company will finally overtake its rivals and create a monopoly (WELFENS, 1995). Traditional monopoly pricing leads to rather inefficient outcomes. The size of the network and the volume of calls are much smaller than required by the economic optimization condition that marginal costs equals price. The social optimum will lead to a deficit if price is set equal to marginal costs (which are lower than average costs). Two-part tariffs are a solution to this problem without state subsidies. Two-part tariffs consist of a unit price p (p equal to marginal costs) and a fixed access charge, payment of which is required for buying consumption units at a constant price p.

Ideally (so-called Ramsey) prices for various telecoms services should be above marginal costs and inversely proportional to the demand elasticities, while taking due account of the budget constraint of the network operator (MITCHELL and VOGELSANG, 1991). But it is doubful that national telecoms firms applied such welfare optimizing price rules. Instead the price structure was distorted by political forces, namely prices were below costs in local telephony but above costs in long distance and international telephony which in effect forced telephone users from the business community to subsidize households. For political reasons uniform prices for telephone rental were applied in western Europe, which again implied massive cross-subsidization, this time in favor of the countryside. This concept of universal services and uniform prices was characteristic of the telecoms network. Universal services were justified on the ground that telecommunications is part of basic infrastructure which government should provide. Moreover, the existence of a state monopoly operator - justified by economies of scale ("natural monopoly problem") - required that the national telecoms operator should offer basic services to everybody. Uniform prices were considered to mimic the law of one price under market competition (OECD, 1995), but in reality there were large international price differentials, especially in transatlantic telephony. After the introduction of long-distance competition in the US in 1984 (divestiture of AT&T leading to seven monopolistic Regional Bell Operating Companies, "Baby Bells", and competition in the inter-LATA long-distance segment) outgoing transatlantic calls from the US were much cheaper than outgoing calls from EU countries (GRAACK, 1997).

2. Changes in the Telecommunications Sector

The first half of the 1990s has witnessed partial privatizations of dominant operators which thereby came increasingly under the pressure of capital markets to achieve a competitive rate of return. Since state-owned telecommunications operators were overstaffed they are responding by laying off employees. At the same time they are moving into upstream and downstream markets as they fear that their traditional main profit source, revenue from network operation, could fall strongly once new network competitors enter the market. The railway companies as well as the electricity companies which are vertically integrated in Europe except for the UK and Scandinavia are also among the main newcomers. With respect to competition in telecommunications this raises a special issue of external vertical integration since these newcomers' core business is a sheltered business activity which could facilitate cross-subsidization and other distortions. The vertical issues in telecommunications concern also the question of internal vertical integration within the sector where driving forces are both the dominant operators as well as national and international newcomers. Vertical integration in essence concerns first information and equipment as inputs into the telecoms network business (Fig. C1), links between the fixed network and mobile telephony as well as links between networks and basic or value-added services (including internet services). However, from a broader view adopted here the issue of vertical integration in telecommunications analytically requires also the inclusion of cable TV and satellite networks because there is partial or indirect overlap with the telecommunications business. Almost all newcomers will face an interconnection problem with the dominant operator which controls the local loop (access to the customer) and thus an essential facility in telecommunications (Fig. C1).

Fig. C1: Vertical Integration in Telecommunications

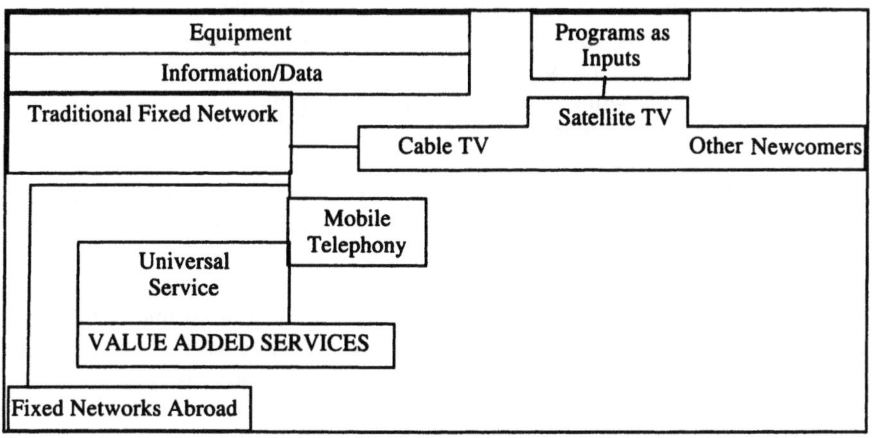

Fig. C2: The Identification of Strategic Potentials

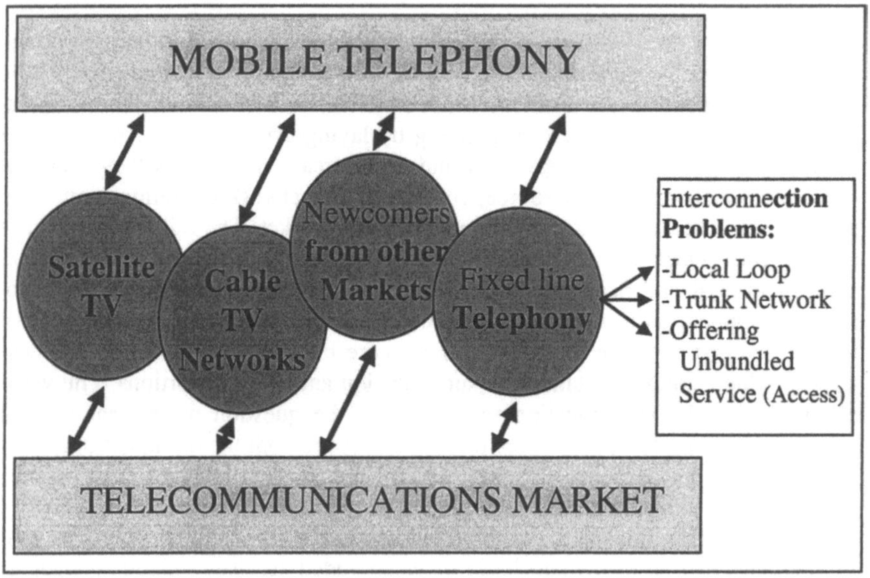

2.1 Basic Regulatory Problems: Theoretical Aspects

A typical problem of network industries is the existence of falling average or marginal costs which leads to the so-called natural monopoly problem: a single firm can offer a certain service at lower costs than if the same amount would be supplied by n independent firms. A crucial problem is that monopoly pricing and monopoly output is much lower than what is efficient from an economic point of view: in the simple case of constant marginal costs k′ output q# is the efficient output characterized by equality of marginal costs k′ and consumers′ marginal willingness to pay. The intersection of the aggregate demand curve DD and the marginal cost curve k′ will determine the efficient output.

With falling average and marginal costs the problem is that marginal cost pricing will not cover costs since average costs k exceed marginal costs k′ – government would have to come up with a subsidy or the monopolist is regulated in a way which allows him to fetch higher prices in the market. If we have different demand curves, DD_I and DD_{II} (for users I and II, respectively), one might raise prices for both groups by a factor h so that there is a uniform price of p_1; the factor h will be chosen by the regulatory authority in such a way so that there is no longer a loss for the company. However, there will be large welfare loss which is shown by the shaded triangles, where the welfare loss is particularly large for the elastic demand curve (DD_I in Fig. C3).

Fig. C3: Consumer Welfare Loss: Basic Problem

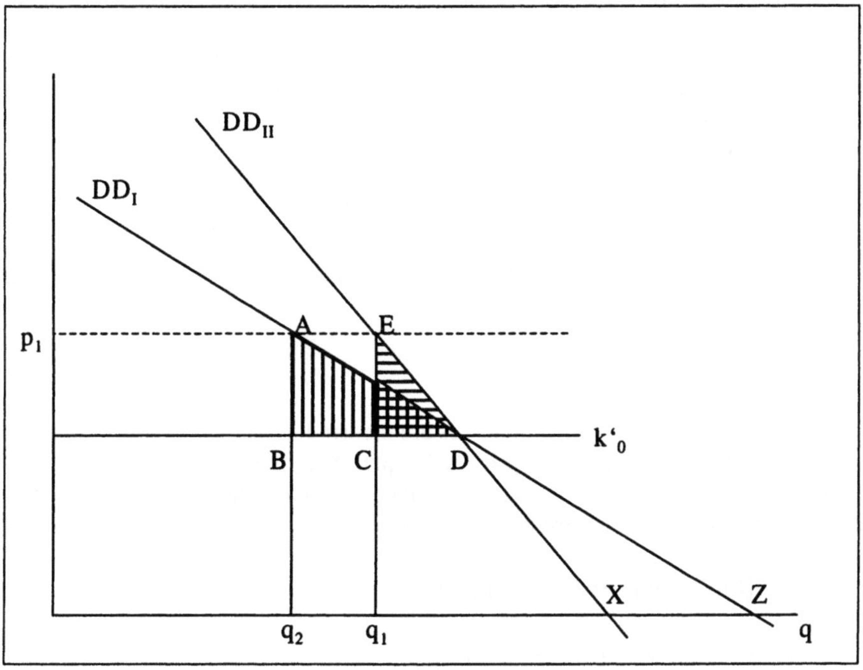

Fig. C4: Consumer Welfare Loss under Ramsey-Pricing

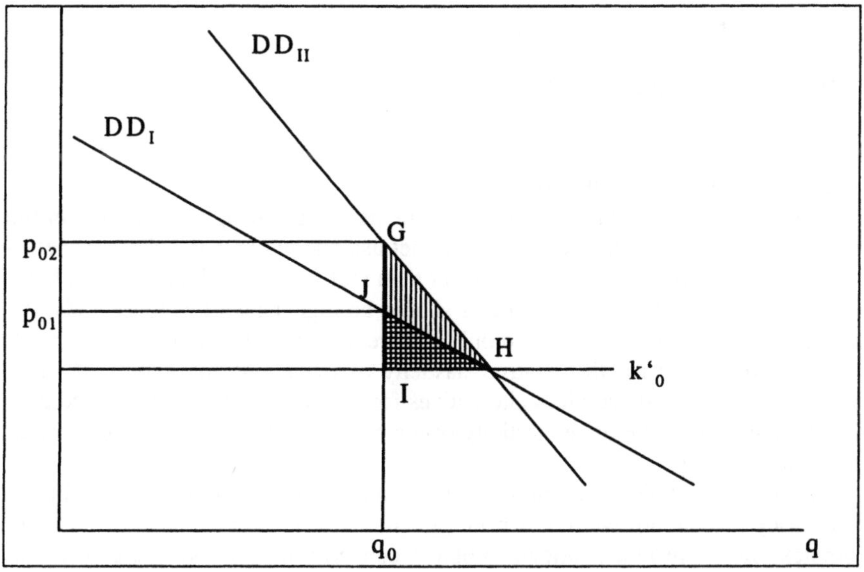

From a theoretical point of view it would obviously be better for public welfare if the regulatory price mark-up for the inelastic demand curve DD_{II} were higher than for the highly elastic demand; note that the elasticity of demand largely depends on the availability of alternative modes of information exchange. An efficient price discrimination formula (with different user groups facing different prices) which minimizes the welfare loss would be the so-called Ramsey prices where the formula for zero cross price elasticity is:

$$(1) \quad (p_1 - k'_1/p_1)\, E_1 = (p_2 - k'_2/p_2)\, E_2,$$

The more general formula (under zero cross price elasticity) for Ramsey pricing is:

$$(1') \quad (p_i - k'_i)/k_i = h/E_i$$

where h is chosen in a way that no losses are incurred; note that in a fully competitive market h is zero, that is the price will equal marginal costs. If marginal costs k' for both users or user groups are identical (k'_o), and demand is isoelastic (that is $q_I = e^{-\alpha p}$, $q_{II} = e^{-\beta p}$), we can restate (1) as follows:

$$(1``) \quad [(p_1 - k'_o)/p_1]\,\alpha = [(p_2 - k'_o)/p_2]\,\beta$$

$$(1``.1) \quad (1 - (k'_o/p_1))\,\alpha = (1 - (k'_o/p_2))\,\beta$$

As p_i will exceed marginal costs we may assume for ease of exposition that k'_o/p_i is rather small, so that the approximation holds $\ln(1-(k'_o/p_i)) \approx -k'_o/p_i$; hence we can logarithmize 1".1 and obtain:

$$(1``.2) \quad \ln \alpha/\beta = k'_o/p_1 - k'_o/p_2$$

If elasticity α exceeds β, making the right hand side of the equation positive, then p_1 must obviously be smaller than p_2.

Hence the price faced by the less elastic demand will be higher than that for the more elastic demand. With respect to internet use one may assume that price elasticity for the business community is lower than that for private households. Therefore regulated internet prices for firms should be higher than for households; the whole argument is, of course, valid only if there is a natural monopoly problem. A second caveat concerns the relative elasticity of firms and households. If firms face more transmission/services alternatives for a given communication need than private households, the price elasticity of the business sector could be lower than that of households.

The above theoretical reasoning applies to any network industry, and problems of Ramsey pricing are well-known in regulations of network industries. A more complex and more recent pricing problem concerns access and interconnection under the condition of opening-up markets cum regulation. The regulatory authority might adopt symmetric price regulation, see for example the UK, Germany, the Netherlands, and many other countries.

2.2 The Competition Problem in Telephony and Foreign Investment

While the natural monopoly argument, based on subadditivity of costs, still holds in part of the telecommunications business, it is clear that without problems of market entry potential competition could discipline the dominant operator. Traditionally high sunk costs - representing barriers to entry - were associated with the local network which also has characteristics of economies of density and scale. However, high historical sunk costs are irrelevant for the competition problem if there exist powerful bypass technologies allowing equal quality of services and lower costs than the dominant operator. With new wireless local loop technologies there are indeed new bypass options although they do not represent a full substitute since fixed network operation offers superior quality. Moreover, if users increasingly demand multimedia applications and other value-added services the unmatched transmission capacity of the fixed network (with fiber glass lines or traditional copper cables in combination with ADSL or HDSL technologies) constitutes a major advantage for fixed network telecoms operators.

The competition problem in telephony basically is as follows: Fixed network newcomers will need - except for cable TV networks - access to the local loop (being an essential facility) which requires interconnection with the dominant operator in those countries where this operator controls the local loop. This is the interconnection problem where the dominant operator can overcharge its rivals which therefore have expansion problems in downstream markets, e.g. European or transatlantic telecoms services for EU customers. Alternatively, a wireless local loop (e.g. based on DECT technology) could be used which implies that mobile telephony is crucial for overall competition in telecommunications, but mobile telephony itself will generate communications which will have to rely on a fixed network, and if this is in the hand of the dominant operator one is back to the crucial interconnection problem. A variant of the problem is found in the US and Finland (ANDERSSON, 1996) where the dominant long distance operator has no direct access to the local loop except via mobile technologies, cable TV – which could be acquired in the US only after the new 1997 telecoms legislation - or explicit interconnection with regional operators. Interconnection is conflict prone and complicated because it has strategic aspects, which can be modeled using game theoretic approaches. Moreover, if the dominant fixed network operator also is active in mobile telephony, which is the standard case in OECD countries, one has to ask whether the interconnection problem does not have two dimensions. First, the dominant operator has an incentive to impose excessive interconnection charges because this supports the expansion of its own mobile subsidiary - an effect which can be reinforced if the private mobile operator has to pay a license fee while the mobile subsidiary of the dominant state operators receives its license for free (an issue between the Commission and Spain and Italy, respectively). Secondly, excessive interconnection charges help the dominant operator to fend off newcomers in telecommunications network operation. From the perspective of a newcomer the dominant operator with its network is a upstream firm for incoming calls and a downstream market for outgoing calls terminating in the fixed network.

The issue could ideally be solved by hiving off the dominant firm's network from telecoms service business. This would make competition for network capacities possible where a weak step towards this model is to allow resale of line capacity and shared leased lines. Given technical aspects and the high capital intensity of telecoms network operation it probably is not optimal to fully disintegrate network operation and telecoms services. However, newcomers should be able to buy services and lease line capacity from the dominant operator in a way which represents unbundling in a very broad sense. By offering only fixed bundles of services the dominant operator could in effect overcharge newcomers.

Unbundling, entry of newcomers and vertical disintegration all imply that contract relations will gain in importance at the expense of hierarchical relations in Europe after 1998. Contract management thus will become more important as will the legal national and international framework in telecommunications. Lawyers should be among the main winners in this process but with gradually rising legal fees and additional legal uncertainties created in due course arms length contacts might be replaced by more firm internal relations as mergers & acquisitions in telecommunications - including vertical integration - will increase in the long term. In the medium term increasing market competition is likely to bring about innovative services, unbundling of business activities and a rising role of contractual relations. From the perspective of OLIVER WILLIAMSON (1985) one could argue that such relations will gain, while hierarchies are temporarily losing in importance.

Privatization together with unbundling and regulation is the key to competition in telecommunications. Ideally, the various telecoms platforms should be separated initially as it was the case in the US as a result of the AT&T divestiture in 1984; or in Finland which for historical reasons had local telecommunications owned by the users, while national and international telephony was organized by Telecom Finland. With liberalization adopted in 1991 the local operators formed Finnet to provide national and international long distance services, too. Telecom Finland lost considerable market shares as Finnet achieved a share of 53% in long distance telephone in 1995, followed by Telecom Finland with 41% (still dominating international telephony) and the newcomer TELIVO - backed by Swedish Telia - which reached 6% (ANDERSSON, 1996). The Finnish example shows that appropriate starting structures and regulation can bring about competition quickly. One may anticipate a similar outcome for the US which removed the market entry restrictions between regional baby bell operators and the long distance operators as well as cable TV companies as of 1997.

Given the theoretical insights and empirical evidence from other countries, including the unsatisfactory British duopoly approach in the period 1984-91 which left BT as a strongly dominant operator, Germany organized the liberalization of the telecoms sector rather poorly (WELFENS/GRAACK, 1996a). Dominant Deutsche Telekom AG (DTAG) was quasi allowed to retain its unprofitable monopolistic West German cable TV network, DTAG only was forced to transform its nationwide cable TV network in several regional networks which then would be sold to private investors – with DTAG keeping a minority stake in the cable TV network. Newcomers in network operation will reduce prospects for DTAG's revenue growth in basic telephony services, and ability of newcomers to use cable TV for the transmission of telecommunications would further undermine the profitability of DTAG.

Since the fields of multimedia and pay-TV are expected to grow rapidly there are impulses for vertical integration with the new growth fields, especially if they require high investment within uncertain legal and market conditions (e.g. in multimedia). While this in effect might help to get competition going it would have been much better to hive off DTAG's cable TV and all or most of its local network operations. Also one could have expected that DTAG would not have been allowed to establish a mobile subsidiary. However, it is clear that in Germany as well as in other EU countries the dominant operators are politically well tied and can influence legislation as well as regulation.

Foreign Direct Investment in Telecommunications

With domestic markets facing competition and given the considerable technology intensity of telecommunications it is clear that the 1990s will witness a wave of foreign direct investment in OECD countries. Telecoms firms will offer services abroad and foreign newcomers will become active which undermines the traditional regulatory policy approach according to which national regulation for a national business was supposed to serve domestic firms and voters. While the telecoms business is internationalizing, regulations are also subject to international influences as both the OECD and the WTO are dealing with the issue of FDI and international services (the US FCC with its new international department is another force stimulating changes). Regulations can affect international services as well as foreign investment in the telecoms sector and beyond.

While in the early 1990s only 1/10 of telecoms traffic was international in major EU countries the privatization and deregulation of telecoms operators plus the ongoing internationalization of the economy (KLEIN/WELFENS, 1992; TILLY/WELFENS, 1996) - i.e. the rising role of multinational companies which require one-stop-shopping - will stimulate the internationalization of the telecoms business. Dominant network operators which face increasing domestic competition will try to earn offsetting new profits from expansion abroad. Since this strategy is relevant for almost all dominant operators in EU countries there will be a mutual invasion of markets which should stimulate market integration as well as mergers & acquisitions in the telecoms business (international aspects of vertical integration).

Since international accounting rates and settlement rates relevant for national operators dealing with each other are much higher than international transmission costs it is attractive for telecoms operators from OECD countries to establish a foreign subsidiary in major partner countries in order to enjoy low interconnection rates abroad and hence to be able to offer lower tariffs for international telecoms calls. For example Swedish Telia established a UK subsidiary with a twin goal: namely to get better interconnection rates from BT which feared that otherwise the Telia subsidiary could expand aggressively in the UK; and to benefit directly from the expanding international and UK telecommunications market (GRAACK, 1997).

2.3 Market Convergence and Distortions from the Media Sector

For technological reasons there is a convergence of telecommunications and media markets. Digitization has eliminated the demarcations between voice, audio, video and data transmission. Cable TV firms thus started to offer telephone services and telephone companies have entered the multimedia and TV market. There are also newcomers from electricity and railway networks as well as mobile network operators. With mobile and satellite technologies as new options to bypass the local fixed network the high historical sunk costs in network building is only partly impairing market entry of newcomers. This holds all the more since private competing cable TV networks which are available on a broad basis in all EU countries, except Greece, Portugal and Italy, offer an alternative bypass option; an exception is Western Germany where the dominant telecoms operator also largely owns the cable TV system.

It is obvious that there are economies of scope in the provision of cable TV and telephony. In addition to the fixed line telephony segment there are communication options in the form of rather expensive mobile and satellite telephony. Satellites indeed are increasingly used for digital direct broadcasting in the US and Europe so that cable TV firms are facing stiff competition. The partial overlap of markets which are regulated in different ways causes specific impulses for integration and internationalization; but it also causes distortions in the telecoms market as we will see below.

Distortions through TV Advertisement

A particularly strange market is TV broadcasting which is widely characterized by advertising-financed programs supplied to households for free. In the end households will pay for the costs of this TV service via higher prices of goods covered by TV advertisement. Given viewers with different preferences and time opportunity costs there is a market for pay per channel or pay per view TV programs so that high income viewers - with a high opportunity cost of time - tend to defect from the free advertising-financed TV program. Demand for free private TV programs is a residual demand which is primarily financed from viewers with inelastic demand buying TV advertised products at higher prices than normal. This leads to a rather poor quality of free private TV (with a program mainly reflecting preferences of smokers, heavy consumer of alcoholic beverages etc.) and a rather small pay TV market which can, however, exist because pay TV networks offer a rich choice of channels (and in the future multimedia services). To the extent that everybody is buying advertised products at higher prices while a certain group is only consuming pay-per-view services there is an implicit cross-subsidization of advertised financed TV, i.e. there are inefficiencies, combined with non-voluntary redistribution among households.

This asymmetric financing of TV - free TV vs. pay TV - stimulates cable TV networks to move into the telephony business. With advertising-financed free TV it is clear that pay TV has a relatively small market in which survival depends on differentiated TV products and additional telecoms service.

Free satellite TV could further undermine the prospects for private pay TV networks. As the satellite TV business is international - indeed an international service business - it would be crucial that adequate rules and regulations would be established in the OECD club as well as in the World Trade Organization. Advertising on TV should be restricted internationally because this causes major economic distortions and even undermines the expansion of an interactive information society - instead a passive TV mentality is promoted and an allocation realized which is similar to that in former socialist eastern Europe where households could enjoy almost free housing resulting in very poor asset management, housing quality and innovation in construction.

It is clear that global rules are mainly set by big influential countries (US, UK, France, Germany, Spain) where TV companies and movie firms can recover a large fraction of production costs so that viewers in other countries - with lower per capita income - can be served at lower differentiated prices. This could e.g. explain attempts by Telefonica de Espana which wants to launch together with Radio Television Espanola, the private broadcasting firm Antena 3 and the Mexican Televisa (plus Unidad Editorial and four public regional broadcasting operators) 40 digital TV channels in spring 1997. In Germany Deutsche Telekom AG is discussing entry in the TV broadcasting market, too, especially after the Kirch group, owning most movie rights for Germany, learned that launching a digital pay TV service in Germany is very difficult without access to Deutsche Telekom's cable TV network. After Deutsche Telekom hived off its cable TV subsidiaries in 2001 – yielding to pressure from the European Commission and taking consequences from high losses in the cable TV business - and sold most of them to foreign investors (DT AG maintained a minority stake in several regional cable TV companies) these investors are eager on the one hand, to invest more in the cable TV network, namely in order to upgrade the network in a way which makes two-way-traffic possible, including internet services, but also telephony and, of course, TV services.

2.4 Theory of Vertical Integration

Vertical integration can save transaction costs that would occur in markets in which separate firms are interacting, and such integration can reduce vertical uncertainty in industry and thereby raise the industry's investment output ratio. While these aspects support the case for vertical integration the latter could also imply allocative inefficiencies which result from internal cross-subsidization, dynamic inefficiency and market power. The issue of vertical integration has been much discussed in the case of the electricity industry where many arguments for integration can be found (SINCLAIR, 1996).

Economic theory suggests five arguments for integrating downstream and upstream activities (for the first four arguments PERRY, 1989; YARROW, 1991; WATERSON, 1993; application to electricity: SINCLAIR, 1996):

(1) reduction in competition;
(2) minimization of transaction costs;
(3) elimination of vertical externalities;
(4) exploitation of economies of scope and scale;
(5) exploitation of network externalities.

(1) Reduction of Competition

Formerly dominant telecoms operators all have incentives to seek vertical integration as a means to reduce competitive pressure. A telecoms network firm could try to foreclose rival firms from market inputs and thereby increase its market share. This happens in an international context through horizontal M&As; because in international telephony networks abroad are a downstream segment for terminating calls so that international M&A activities have a downstream aspect. This is of particular importance since international interconnection typically is a field of monopolistic interconnection pricing which thus restrict the expansion of international telecommunication calls; moreover, arm´s lenght transactions via markets will involve highly complex and lengthy negotiations on international interconnection – ie there a high market transaction costs which imply that multinational investment strategies which replace market relations by firm-internal relations (supposedly at lower transaction costs) should be rather popular in telecommunications once such strategies become feasible. Once privatization in the telecommunications sector opens new opportunities for investment abroad one therefore should expect a wave of foreign investment which will lead to the phenomenon of multinational companies (MNCs) in the telecommunication sector. Networks abroad are particularly lucrative if there are many outgoing calls from the network of the MNC´s parent company country or if many calls from foreign markets – with the presence of a subsidiary network operator – are the source of outgoing calls which terminate in the parent company country. To some extent this could be the reason behind the aggressive investments of Telefonica in Latin America. In 1998 Telefonica had 23.5 million customers in Latin America compared to 21.6 million in Spain. Assuming that traditional and new trade plus investment links – outside telecommunications – create high growth rates of voice and data traffic between Spain and Latin America Teleconica´s Latin American expansion strategy should pay off in terms of a higher market capitalization. However, one may wonder whether Telefonica can survive as an independent European operator. Telefonica already has formed some strategic alliance with WorldCom-MCI, and the chairman of Telefonica was on the board of MCI. However, alliances generally have turned out to be rather unstable in telecommunications. The rapidly rising competition stimulates international rivalry and oligopolistic price wars.

Typical upstream and downstream integrations have occurred in cable TV. Cable TV networks are linking up with the movie industry in order to impair market entry of newcomers. Integrating downstream activities - e.g. multimedia and TV emission - is in the interest of telecommunication network operators who want to raise the volume of network use and to reduce uncertainty in their investment planning. Regulation can be a factor encouraging anticompetitive strategies of integrated firms (YARROW, 1994; VICKERS, 1995). Telecom firms facing price caps and broad universal service obligations will try to obtain offsetting high profits from other activities. If unnecessary comprehensive universal service obligations are imposed upon the dominant operator he/she will be encouraged to proceed with vertical integration - especially since after privatization capital markets expect a competitive return on investment. By achieving high profit rates in unregulated activities the dominant operator can mitigate the profit-depressing effect of excessive regulation. Since the policy course of the regulator is uncertain in many countries dominant privatized operators will try to pursue a vertical integration strategy as a kind of insurance strategy; namely, if the regulator should impose excessive price cap regulation unregulated upstream or downstream fields will have to generate extra profits. Since regulation is transitorily increasing in the first stage of telecoms liberalization there could be a strong incentive for vertical integration in the early stage of deregulation.

(2) Minimization of Transaction Costs

Transaction costs will be considerable for movie and TV companies which have to negotiate access to broadcasting and network capacities. The problem of imperfect contracts in a technologically dynamic network environment suggests that vertical integration has major advantages in the TV and telecoms business; WILLIAMSON's (1985) list of factors encouraging vertical integration, i.e. replacing external governance structures and markets by internal governance, fully applies in the TV/telecoms network segment: there is a high specificity of investment, transactions are occurring frequently and there is considerable uncertainty about future market conditions. Since telecoms services can be offered via TV networks this tendency will indirectly affect the telecoms markets. High asset specificity leads to the "hold-up" problem. Once the cable TV company has undertaken investment in specific technology and equipment the other party is in a superior bargaining condition to renegotiate existing contracts and to appropriate quasi-rents (excess profit in comparison with putting resources invested to a different use).

Mobile telephony which partly relies on the dominant operator's fixed network is a case where vertical integration pays off from the point of view of the fixed network operator since such integration implies lower negotiation costs and legal uncertainties which are observed in pure contractual relations. Similarly, the local loop and the long distance business could benefit from vertical integration. However, the static gains might be offset by dynamics efficiency losses - i.e. insufficient innovation and product differentiation - which would not occur under vertical separation.

Since deregulation and privatization of telecoms will stimulate foreign direct investment in telecommunications - reflecting attempts to obtain offsetting network business (and better interconnection rates) abroad in a period where newcomers are entering the domestic market - one may anticipate bigger European telecoms operators to emerge soon. With internet providers facing new big telecoms operators they will fear increasing price uncertainty in the future. This could encourage service providers to vertically integrate in part of the EU market in order to have better information about true network costs. HART and TIROLE (1990) and BOLTON and WINSTON (1991) argued in their models that vertical integration and market foreclosure can be in the interest of firms. ORDOVER, SALONER and SALOP (1990) have emphasized that vertical merger can facilitate collusion at the upstream stage, raising intermediate goods prices and the costs of downstream competitors and thereby leading to higher final goods prices. The ability to foreclose downstream competitors from the supply of inputs (charging a higher price to non-integrated downstream rivals and thus raising the own final goods price) will be constrained by the presence of other upstream firms. The motivation of monopolistic local or national network operators to integrate vertically with competitive activity levels - e.g. multimedia input or movie firms - is often rather strong.

Voice telephony and telecoms services will be largely liberalized by 1998 except for universal services so that former telecommunication monopoly operators which anticipate increasing network competition are encouraged both to pursue internationalization and vertical integration; the latter means expansion of value-added services. In most continental EU countries value-added services account for a rather small share of revenues but in liberalized Scandinavian countries the share is much higher.

(3) Elimination of Vertical Externalities

Vertical commercial externalities play a role in telecommunications. Such externalities will occur if firms have market power. If network operation and service provision are separated upstream or downstream firms might underinvest from the point of view of the industry as a whole. For example investment in advertising for telecoms universal services use could generate additional demand for value-added services so that the network operators have an incentive to integrate vertically. Moreover, with more mobile phone users the demand for fixed network capacity will increase, while opportunities for full capacity utilization of switches and lines in the fixed segment will improve.

(4) Exploitation of Economies of Scope and Scale

Joint network management could reduce costs; value-added services could be offered at discount prices for time slots with low telephony demand. Appropriate contracts of the network operator with prices differentiated during the 24 hour cycle and price differentiation of service providers could achieve this result as well. Vertical integration of cable TV (network operation) and telecoms services is, however, important for cable operators since offering rather cheap telecoms services is a means to reduce the churn rate in the cable TV business.

Since marginal telecommunications costs are close to zero in off-peak periods telecoms operators are interested in raising penetration rates and generating additional data communications. This will help to exploit economies of scale which are extremely relevant for glass fiber cables; recent seas cable have contributed to sharply falling transatlantic telecoms rates (the new Flag Cable with four glass fibers can carry up to 600 000 simultaneous conversations).

(5) Exploitation of Network Externalities

Network externalities are important in telecoms. Demand-side network effects mean e.g. that the marginal utility of having telephone access increases with the number of other users. This requires a critical minimum number of users: with more users to communicate with it becomes more interesting to have a telephone (ALLEN, 1988; WELFENS, 1995). In vertically integrated telecoms networks such externalities can occur at different layers. Access to basic telephony is one layer (as in the above example) in countries with a low penetration rate; in countries with a high penetration rate positive network externalities will rather occur at the value-added services level or between fixed and mobile telephony. Fax services or multimedia services will have network externalities in the early expansion stage. For example if a critical minimum number of colleagues is using multimedia services or video-conferencing at home or in the office it becomes crucial for everybody in the firm to also use multimedia services or video conferencing. A telecoms operator will have an incentive to provide cross-subsidized multimedia access - including compatible software products - as long as this helps to achieve the minimum critical number of users.

AT&T launched its own subsidized internet services with remarkable success in 1995. From a competition point of view there is no need to restrict such vertical integration as long as other internet providers have access at nondiscriminatory conditions to the AT&T network. In Germany the Deutsche Telekom AG has also launched its own multimedia service, T-Online, which is rather successful. However, other internet providers claim, that Deutsche Telekom AG does not give fair access to its network. By creating marginally worse quality of services for outside internet providers Deutsche Telekom AG can hope to generate for its own network the desired critical mass and a maximum market or revenue share.

As long as network operation and telecoms services are not split up by the regulator or by market forces there will be the issue of universal services offered at uniform prices which will be mainly provided by the dominant operator. With several operators the question arises whether one should not switch to a concept of differentiated universal service obligations which would mean that operators must provide network access and services - however, not at identical prices to all users; rather at predetermined prices for different groups of users requesting access and services. If basic services were unbundled in the new competitive EU environment one could even create new user options in the sense that various services could be bought from different operators. Depending on the price accepted and the technological development (quality of service) households or firms could chose flexibly among competing operators. This would stimulate efficiency and technological progress.

Fig. C5: Reasons for Vertical Integration

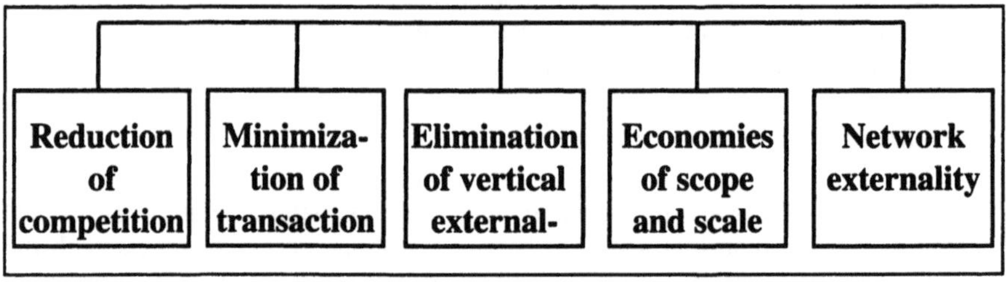

2.5 Special Aspects of Telecoms and Information Markets

For well-known reasons information markets are rather imperfect. Those who supply information have to reveal part of the information package in order to allow the demand side to assess the value of information. Under such circumstances the producer of information will never be able to fully recover costs so that provision of information will be suboptimal in a pure competitive setting. One solution to the problem is that information providers are encouraged to build up reputation which will allow them to raise prices and thereby to expand the information business. Government could stimulate such a process by establishing independent audited ratings for information providers. The aim would be to minimize the disappointment risk of information users, namely by directing them to specialized information providers and to those communication firms which are known to provide reliable and useful information at an adequate price. This in turn would reduce pressure for vertical integration. This will help to create a true information business in which information has a positive price. Rating agencies in financial markets are an example for the virtue of rating, although the rating market itself is not fully competitive since major misperceptions have occasionally occurred (e.g. Mexico crisis 1994/95) and market exit and entry were very rare over the past decades. However, the rating approach is a promising avenue to get the information society going.

The alternative is to allow vertical integration in the information business so that integrated network operators can offer both network transmission and information services. A network operator has poor incentives to create a selective information market with specialized providers since its main interest will be to generate demand for network capacity in the medium term. This interest is served best by creating a mass market in which users bear the cost of information acquisition by surfing in the net over extended time periods (as soon as a critical mass has been achieved a strategy of price differentiation makes more sense). The average cost of finding information then consists of internet access price plus opportunity costs of surfing time; with insufficient differentiation and rating of the information supply side the demand for browsing software will increase, but even powerful browsing software can hardly remedy the inefficiency of the information supply side. A flat rate for internet services - typical in the early 1990s in the US and Europe - encourages hours-long surfing. The flat rate tariff is likely to end soon in the US as local network operators in the US start charging the internet providers who for a long time were free riders on the US national science communication network. The traditional digital telecoms switches used by local network operators are not ideal for internet providers as they are built for passing short voice messages. Internet providers might themselves consider to acquire local network operators or to obtain regulation which allows them to install their own switches within the switching systems of regional operators.

For a true information market to develop in OECD countries it is crucial that information services are not simply bundled with network transmission. Moreover, it would be useful to develop a politically independent - possibly subsidized - information rating system. In order to avoid discrimination in information provision information providers with a certain minimum rating should have a right to access to public switched telephony networks.

2.6 Regulations of Telecommunications and Internet Expansion

The telecommunications sector has traditionally been considered to present natural monopoly problems associated with declining marginal and average costs. In the US, AT&T ("Ma Bell") used to have a private quasi-monopoly until the dismemberment of 1984 when long distance competition was established. The most crucial problem of regulation concerns local telephony. The US only moved to conditional market opening up in this field after the introduction of a new law in 1996: Local network operators (the Baby Bells) were allowed to enter the long distance market provided that they opened up the local telephone market. However, even five years after this liberalization, newcomers have less than 5% market share in local telephony. AT&T's strategy to enter the local telephony market via a cable subsidiary has largely failed. Regional baby bell companies merged in the late 1990s and have at the same time become successful mobile operators. With international telephone prices falling quickly since the EU liberalization of 1998 – and liberalization in Asia and Latin America – prospects for profits in international telecommunications are modest. Against this background the regional baby bells are in a relatively strong position since access to the telephone network is a truly resistant bottleneck which cannot be easily bypassed as long as the facility for powerful (3G) mobile telecommunication technology has not been rolled out. DSL technology will allow the upgrading of the fixed network so that it will remain a nation-wide powerful network for communication.

There is no doubt that EU liberalization in 1998 – and the UK liberalization after 1984 - has strongly stimulated competition in telecommunications in Western Europe: Prices have fallen steeply– particularly in long distance and international telecommunications where competitive pressure and regulatory requirements, that is price caps, have helped to bring prices down. However, prices for access and for local telephony have only modestly fallen, in Germany for instance there was even a price increase in 2001; Deutsche Telekom, the former national operator, has maintained a market share of 95% in the access market and local telephony; value-added newcomers accounted for only about 12% of the overall telecommunications market revenue in 2000.

Product Bundling as a Regulatory Challenge

Developments in the US and Western Europe clearly point to the specific problems encountered in the areas of local loop and telephone access. There is indeed the problem of a transfer of monopoly power from the local telephone market to other markets – such transfer of monopoly power has always been an area of concern for the FCC in the US, but also for the European Commission and national regulatory authorities in Western Europe. Regulatory authorities in telecommunications have typically focused on local access and interconnection plus local telephony; in Europe also on long distance and international calls where price caps on the latter field might be expected to be phased out as mutual invasion of national markets by foreign former monopoly operators within Europe has, since 1998, strongly intensified competition in this segment.

It has been noted that following a first stage of market opening – when newcomers gain customers via call-by-call or on a preselection basis – there is a tendency for aggressive product bundling (combining in various ways local telephony, long distance, international calls, internet plus mobile calling plans) which amounts to reinforcing the dominant position of the former monopoly operator; the dominant operator can pursue predatory pricing. Bundling has reduced market entry dynamics because newcomers are forced to come up with much higher investments as they need to cover several markets which are highly interdependent due to product bundling.

For simplicity we can distinguish local access and local telephony which both stand for a resistant monopoly – most crucial is local access since it represents an essential facility needed for other services. In addition to the infrastructure pillar and the voice telephony pillar there is the value-added services pillar and the equipment pillar which constitute the telecommunications market. Bundling can occur across all four pillars.

Whether product bundling is normal product bundling as would occur in a fully competitive market is rather doubtful. The fact that the dominant telecommunications operator is the leader in product bundling raises some unpleasant doubts: Should the regulatory authority which conducts asymmetric regulation in order to offset the unavoidable early dominance of the former monopoly operator allow all kinds of bundling, in particular cases where regulated products and unregulated products are combined – which often effectively undermines regulation. From an economic perspective, bundling of products makes demand curves in all concerned markets less price elastic, leading to higher profit margins. Aggressive product bundling weakens competition by raising barriers to entry in the telecommunication services market. At the same time it makes substitution on the consumer/user side more difficult.

Another problem with competition concerns cross-subsidization on the side of the dominant operator which could also distort competition; if the former monopoly operator has high profit margins in some segments which then are used to cross-subsidize expansion of new services (eg DSL in the access market) there is a risk that market power be reinforced. Weakening competition in turn would reduce the prospects for static and dynamic efficiency gains.

Fig. C6: Problems of Product Bundling in Digital Services (bundling of products on the side of ex-monopolist reduces price elasticity of demand in all telecoms markets)

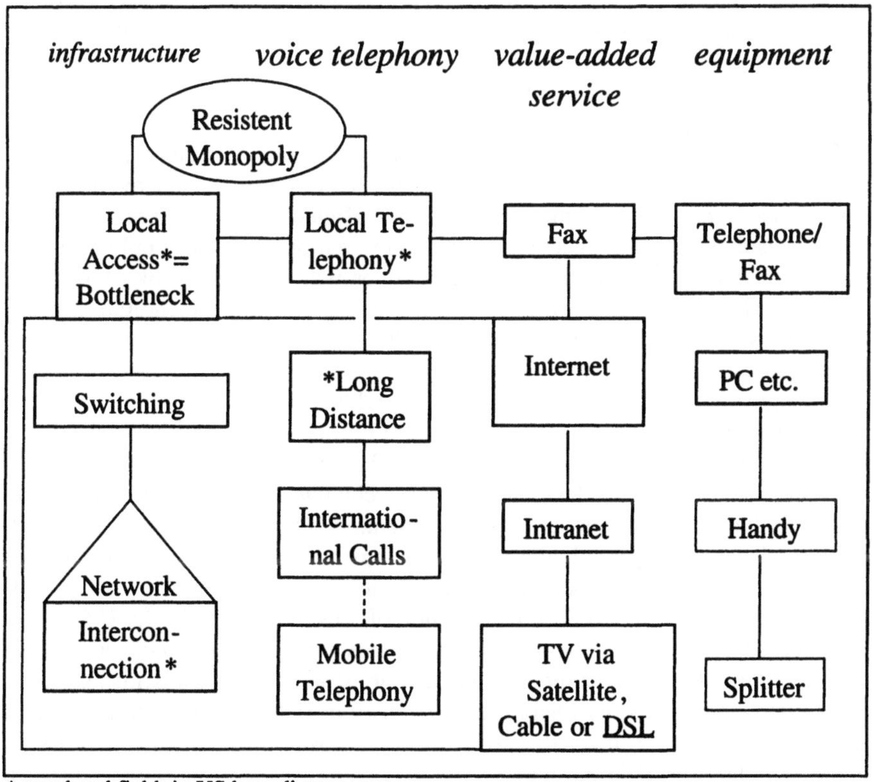

* regulated field; in US long distance not
DSL=digital subscriber line

3. Traditional and New View of Telecommunications

Traditionally EU countries coped with the problems of falling average costs - the case of a natural monopoly - by opting for a state owned telecoms monopoly. While there are some arguments for economies of scale and economies of density in fixed line network operation there certainly is no natural monopoly in the field of telecoms services. From an economic perspective the optimal way to ensure efficiency would be strict unbundling between network operation and telecoms services on the one hand and open market entry for newcomers in network operation on the other hand. With traditional technologies newcomers faced the problem that the local network in European countries - except for Finland - was in the hand of the dominant state-owned operator so that complex interconnection agreements had to be reached. However, progress in mobile telephony allows a wireless bypass solution where operators using the new digital transmission technologies (GSM, CDMA or DECT) face the problem of the scarcity of radio frequencies. For this reason alone there has to be some form of government regulation even in the future.

The telecoms sector is characterized by (1) globalization of business, (2) high technology dynamics and (3) high capital intensity. Large capital requirements make a stable regulatory framework necessary. High sunk costs - specific investment costs that cannot be recovered when going out of business - imply a market entry problem: the incumbent network operator will disregard sunk investment costs (which include part of fixed investment and advertising expenditures) when newcomers want to enter the market. For the newcomer all costs are non-sunk prior to market entry so that newcomers can enter only if they command sufficient technological or special cost advantages (e.g. rights of way).

As regards access to the network in western Europe capital costs are indeed falling since even in rural areas households can be connected to the terrestrial network at small costs if wireless transmission techniques (radio in the loop) is applied. Pricing of the national network operator was a traditionally sensitive political field in Europe. Urban users and the business community in particular in effect cross-subsidized households, especially in rural areas. Prices for all groups can, however, fall - as the 30% real decline of telecoms prices in the UK in the past decade suggest - if competition and hence productivity growth is strong. Competition in the UK was limited to a duopoly strategy until the British government switched to a broader liberalization strategy in 1991. The British lead in EU telecoms deregulation fostered British operators position as foreign investors in liberalizing latecomers in Europe.

The EU Commission has launched several studies and initiatives to stimulate competition in an industry that is crucial for the advanced OECD countries which increasingly will become information-based economies: most notable are the Commission's White Paper on 'Growth, Competitiveness and Employment' (COM(93) 700 final) and the Bangemann Group report on 'Europe and the Global Information Society'. The report suggests that Member States liberalize the telecommunications sector by "opening up to competition infrastructures and services still in the monopoly area". Finally, the Green Paper on the 'Liberalisation of Telecommunications Infrastructure and Cable Television Networks' (Part I and II) envisages a broader range of concrete steps towards liberalization.

The Green Paper (II, p. I) raised key questions for telecommuincations: (1) *How can universal services be developed? How much does it cost and who should pay for it? What should be the future framework for interconnection and interoperability? How will telecoms networks be licensed? What restrictions can legitimately be placed on licence numbers and what conditions be attached to licenses? How can a fair competitive environment be ensured? How can infrastructure competition contribute to the challenge of employment? How should Europe address the shift in employment in telecommunications? How can Europe ensure comparable and effective access to global markets? What are the broader societal effects of the Information Society and how can the regulatory framework address these?*

The EU will have with few exceptions a liberalized market for network operation and services by 1998 and there are prospects for an enormous increase in competition (WELFENS and GRAACK, 1996). Since the EU's Green Paper (II) suggests that regulation should largely remain on the national level, the new technological developments ultimately require national telecoms regulation to effectively focus both on the traditional telecoms network operation, the cable TV sector and the electricity sector. The latter sectors were disregarded in the Green Paper. In addition to this railway operators, which all have an advanced internal communication network could enter the market. The same applies to internal networks which were set up by banks and multinational companies in the EU for transmitting data. As a matter of fact digitization in principle also allows transmission of voice and other "bits of information" for third parties. Even in small countries such as Sweden or Finland several competing network operators can profitably coexist.

3.1 Vertical Integration and Political Economy of Liberalization Policy

In a competitive market economy a vertically integrated sector should be an exception not only because vertical integration could create serious inefficiencies but also because competition can be achieved in such a setting only with adequate regulation or vertical disintegration. However, a vertically integrated sector will consists of rather large and therefore influential firms which will try to achieve regulation that in effect further impairs market entry.

Regulatory capture is a typical problem of regulation emphasized in the literature. With liberalized EU markets operators will become bigger and more international; indeed with big vertically integrated operators facing competing national telecoms operators the opportunities for regulatory capture might be rather large in the EU in the future. Analysis of the political economy of the telecoms liberalization process shows (WELFENS/GRAACK, 1996b) that once a critical level of privatization or internationalization plus deregulation has been reached the competition process will transitorily accelerate. If the critical minimum effort is not made, the public monopoly is simply transformed into a private monopoly.

Vertical disintegration prior to privatization of a state monopoly is a convincing concept from a long term economic perspective which suggests that getting the initial starting conditions and regulation right will result in an evolutionary competition process which will bring about optimum market structures, i.e. firms that are efficient in both a static and a dynamic sense. However, political rationality is a short term concept and rather favors maximizing short term privatization revenues - this in effect rules out dismemberment of the dominant operator (remember that AT&T divestiture occurred in the context of an antitrust case!).

3.2 Implications for Deregulation

Accounting separation should be required from integrated firms in order to minimize distorting cross-subsidization. This indeed partly is the case in the UK and Germany. However, in reality accounting separation still allows firms to manipulate accounting figures for distinct activities.

Quite complicated is the interconnection issue, although the UK and New Zealand have developed basic rules that seem to solve the conflict. If interconnection issues are taken to the Courts it takes many years to sort out the problems, a clear disadvantage for newcomers so that the New Zealand system which initially intended to have a system without specific telecoms regulations turned out to be not ideal for competition.

Universal service obligations should be rather limited if anticompetitive vertical integration impulses are to be minimized. One might indeed introduce the concept of differentiated universal service obligations. Regulations that want to ensure efficiency should prohibit free telephony based on advertising which already is offered in Sweden - using free phone the user has to listen every ten seconds to advertisements. Such a marketing concept for fixed networks could become popular as an answer to innovative cheap internet telephony in the long run. Here we face a matter of principle, and indeed in a market economy there are very few goods which should have a zero price unless one would like to have serious distortions. If poor households are to obtain favored access to telephony this can be arranged by specific transfers. With tight public budget constraints the pressure to allow free phone services could indeed increase; moreover, with international satellite services it might not be feasible to prohibit advertisement during (satellite) phone conversations. However, the government can introduce a high advertising tax which could be an adequate means to minimize advertising both in the case of free phone and free TV programs.

The telecoms sector is only one of several public utilities. From a pro-competitive point of view policy credibility in network industries will be reinforced if regulatory agencies are politically independent and adopt similar courses in the field of vertical disintegration.

Government could support vertical disintegration in the course of privatizing the telecoms business. In Germany this was not considered seriously so that

Deutsche Telekom AG was allowed to maintain its quasi-monopoly in Western Germany's cable TV network. Private cable TV companies and potential newcomers can try to push for divestiture by either going to the court (invoking Article 90-92 of the EC treaty) or by taking the issue to the Commission. The Commission has poor prospects in achieving divestiture going unless the dominant operator needs the Commission's consent in other areas (e.g. the case of Spanish Telefonica which needed clearing for the multinational Unisource consortium but got this only at the price of effectively waiving the extra transition period for Spain's network liberalization). Vertical integration could be a preferred way to acquire market power in the case of small and rather unprofitable telecoms operators. It seems that the network operators will in any case be the core around which vertical integration takes place since they control the gateway to the final consumer.

Another problem could be external monopolization which occurs if foreign investors from country A invest in country B's competing operators - but later A's firms merge. This creates the problem of secondary monopolization. International operators could pick that EU country for vertical integration which is most liberal or most desperate to privatize its uncompetitive telecoms company.

The issue of advertising-financed TV is certainly very critical in both the US and Europe. First, TV is politically sensitive so that governments will be reluctant to undertake anything which is contrary to the interest of mass TV stations. Secondly, the international nature of direct satellite broadcasting renders an isolated national solution difficult.

A pragmatic strategy of economic policy could be to adopt asymmetric integration policies in the sense that firms from a competitive upstream or downstream sector will be allowed to integrate vertically to some extent, while dominant operators will not be allowed to aggressively pursue vertical integration. Under normal profit conditions the monopoly segment is, however, poised to be the more aggressive and financially powerful raider. This leads to the requirement that dominating (inherited) telecoms positions are eliminated in the EU which in turn requires the creation of greater markets and vertical divestiture. From this perspective the Dutch government took a prudent decision in 1996 when KPN, the telecoms operator, was forced to reduce its stake in one of the largest cable TV stations to a minority stake of 20%. It remains to be seen whether other EU countries will adopt a similar approach and create an initially level playing field for the former national operator and alternative networks. With appropriate starting conditions evolutionary development in liberalized markets certainly will lead to some degree of vertical integration, but this integration should not - except for full integration - be prohibited by regulation as long as it is welfare increasing.

Minimum information requirements on quality and price of services would be useful to enhance market transparency and achieve efficient universal services. Market-driven pressure to regularly publish a broader set of performance data could also result from ongoing privatization, i.e. the impulses from capital markets. With such markets gaining importance in Europe there will be pressure to avoid inefficient structures of firms. However, there will also be incentives for operators to seek protection against potential hostile foreign takeovers which means that firms could try to move into fields in which restrictions for foreign investment exist.

4. Competition in Network Industries

Competition is crucial for efficient allocation of resources and thus for economic growth as well as for structural change. For customers competition means to have a choice and to be able to compare alternatives which motivates firms to produce at the lowest possible costs and to develop product innovations and process innovations.

Major economic benefits in terms of efficiency gains can only be expected if competition is combined with private (or privatized) companies. It is true that private firms suffer from inefficiency problems, especially in the presence of more or less competent managers: The board of managers might not employ resources in an optimal way which is in accord with the preferences of stockholders and owners, respectively. However, this so-called principal-agent problem is even stronger in state owned firms, where citizens and taxpayers (the ultimate owners), respectively, have even much poorer ownership rights, while government bureaucrats - nominally having ownership rights - are weakly motivated to supervise the management of state firms and also tend to interfere for political reasons with normal business. Private firms whose managers have to achieve a normal rate or return under the pressure of the capital market will face stronger incentives for efficiency. This holds in particular, if managers are rewarded in accordance with their efforts and the development of shareholder value, where the latter is achieved via stock options for managers.

Competition is important for structural change. Firms and sectors, respectively, which record rates of return above normal market rates will expand their capital basis, and newcomers will be motivated to enter the respective markets. By contrast, firms and sectors with rates of return below normal market rates will shrink. This holds only if government is not preventing this process by subsidies for ailing industries or external protectionism (e.g. tariffs). By stimulating expansion of firms with high productivity growth and hence high profit rates and by encouraging a shrinking of obsolete industries competition helps to achieve economic growth in a market economy.

Government has the fundamental task to protect competition and to get competition going in monopolistic industries. Moreover, government has to realize the rule of law because such a regime is the only way under which contracts - the fundament of a market economy - will be used by market participants in a wide range. Government can support economic growth also by subsidizing research and development, by modernizing the education system and by a stable monetary policy.

Generally, competition means that there are several firms in the relevant market. The radius of the relevant market is determined by normal conditions under which suppliers and customers will meet; falling transportation and information costs clearly enlarge the relevant market radius (for bread the relevant market is local, e.g. a town or city, the market for holiday trips could encompass the whole nation or a whole continent). If government does not protect competition there will be a tendency towards eliminating competition in the long run; with each business cycle trough and each sectoral crisis there will be tendencies in industry to consolidate, where the formation of cartels and mergers & acquisitions are two typical ways to reduce competition. Therefore in most OECD countries there are - disregarding certain exemptions, e.g. in R&D or standardization - strict anti-cartel laws as well as a merger control by authorities.

A special case of competition policy are the so-called natural monopolies, i.e. industries with falling average costs which imply cost advantages for big firms and ultimately an endogenous incentive for monopolization via mergers and acquisitions. Monopoly means static and dynamic inefficiency. The consequences of monopoly typically are excessive prices, slow rates of technological progress, output below the level of competition and poor service quality. For decades governments in OECD countries sought mainly two ways to remedy the contradiction between monopolistic behavior and the efficiency interests of consumers. In telecommunications the supply side in natural monopolies was organized as a single state-owned firm (Western Europe) or as a private regulated monopoly (USA).

According to theoretical analysis of BAUMOL/WILLIG (1981) under certain conditions industries with falling average costs - e.g. telecommunications, electricity and the railways system - can be characterized by effective competition, namely potential competition: the credible threat of market entry of potential newcomers could be sufficient to impose competition discipline on the incumbent monopolist (high monopoly prices naturally attract newcomers). This, however, will hold only if the so-called sunk costs are low as is emphasized by the BAUMOL/WILLIG approach of contestable markets. One has therefore to distinguish between potential competition and actual competition; a further refinement (Fig. C7) will distinguish between nontradable goods and services and tradable goods which are exposed to world market competition via trade. Nontradable goods and services have an international dimension in competition only to the extent that mobile foreign investors have access to markets (and that customers are mobile internationally). Foreign direct investment is crucial for efficiency gains if the industry is characterized by a high R&D intensity and if the role of potential domestic newcomers is limited.

Competition typically is very intensive in the tradables market where suppliers are subject to world market competition. In the nontradables market competition is less intense. The services sectors largely is a nontradables market, although there are some fields in which high international mobility of users or suppliers effectively creates a very competitive international market. Modern telecommunications can significantly increase the mobility of various services, e.g. financial services. It can also facilitate the creation of flexible networks in industry because with modern telecommunications even large firms can be organized in a rather decentralized manner.

The presence of foreign investors can intensify competition. True competition could, however, be seriously impaired if only one dominant foreign investor is allowed to enter the market. Such a tendency sometimes is encouraged by the host government's interest in getting maximum short term proceeds from selling assets to an exclusive foreign investor which - anticipating a future monopoly - position will bid more than the average foreign investor in a situation with several bidders. One must emphasize that short-run revenue maximization of government means to disregard the long term benefits from competition and the presence of several foreign investors, namely sustained higher investment, growth and tax revenues. If the political decision horizon is very short, e.g. because of political instability, short term revenue maximization might become a principle of government policy.

Fig. C7: Competition in Industry

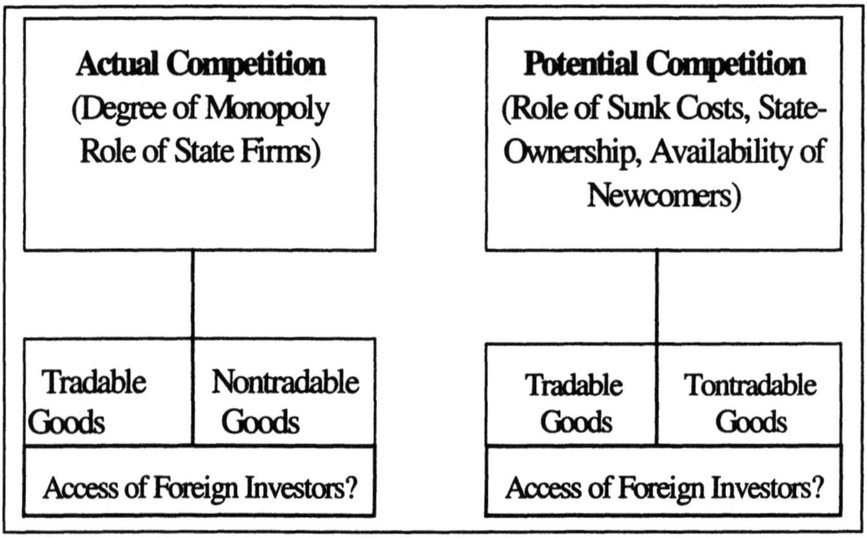

Sunk costs are industry specific investment costs that cannot be recovered after market exit. While the market entrant will consider all costs as variable prior to market entry the incumbent firm will disregard sunk costs when it comes to strategic aggressive pricing behavior; i.e. set prices so low as to discourage the potential newcomer from market entry. If sunk costs (specific investment costs, R&D costs, marketing costs) are high - as in the case of network industries with high investment into terrestrial networks - potential competition will be weak unless there are significant technological innovations which allow to bypass existing networks at low costs. New wireless technology is available since the 1980s.

4.1 Network Theory, Technological Progress and Competition

Telecommunication infrastructure is important in economic terms for the following five reasons:

1. It generates positive income and employment effects as the infrastructure is built,
2. It increases the economy's overall fixed capital stock, raising the productivity of complementary factors and potential output,
3. It allows the provision of services which are crucial inputs for many sectors,
4. It generates indirect pro-competitive effects as infrastructure expansion increases the size of markets, provided that there are no recurrent congestion,
5. It helps increase the provision of services by competing service providers who require control systems necessary for optimal hazard-free capacity utilization - control systems can be organized by the government as in the case of police services in road traffic, by private monopolist as in the case of air transportation or by vertical integration as in telecom, electricity and railway systems.

Modernizing network industries is crucial for economic prosperity as it allows companies and governments to cope with rising trade and traffic in the new Europe and hence with a globalizing economy. Moreover in this field, private sector activities have become more important for two reasons : several traditional arguments for exclusive government provision of infrastructure seem no longer valid and tight government budget constraints in almost all industrialized countries are stimulating the search for new ways of organizing network industries. Competition in network industries can take various forms, ranging from regulated monopoly to competition for the market (network), competition in the net (separation of network and competitive services), and finally full competition of integrated networks (Fig. C1). The late 90s announced the end of regulated monopoly in the field of telecommunication for all OECD countries although the US and the UK had already liberalized their markets by 1984.

Network operation, control and services - represented as three distinct infrastructure layers by KNIEPS (1997) and illustrated in Fig. C2 - can, in principle, be vertically separated although national monopoly operators tend to organize these activities jointly. This is because network industries are highly capital-intensive, and it would be in the interest of the network-owner to ensure profitability of investment and provision of profitable services via vertical integration. This however often impairs market-access. In the telecommunication market, technical characteristics should combine network control and operating facilities.

Infrastructure networks are often integrated systems which offer commercial services over space. Typically, one may expect some basic public interest in the general availability of such services. However, this gives rise to a universally common service problem which involves either a basic cross-subsidization problem or an explicit deficit-financing problem. The regulations imposed are designed to ensure appropriate public service. However, the positive theory of economic regulation (VON WEIZSÄCKER, 1982; KNIEPS, 1988) argues that the regulatee has an incentive to create a monopolistic situation and to engage in rent-seeking activities, i.e. lobbying-activities, financed out of monopoly profits and therefore regulations are designed to encourage monopolistic situations by creating market-entry barriers. This raises the issue of whether regulatory institutions should be independent or not from both the industry and the political system.

Fig. C8: Competition in Network Industries

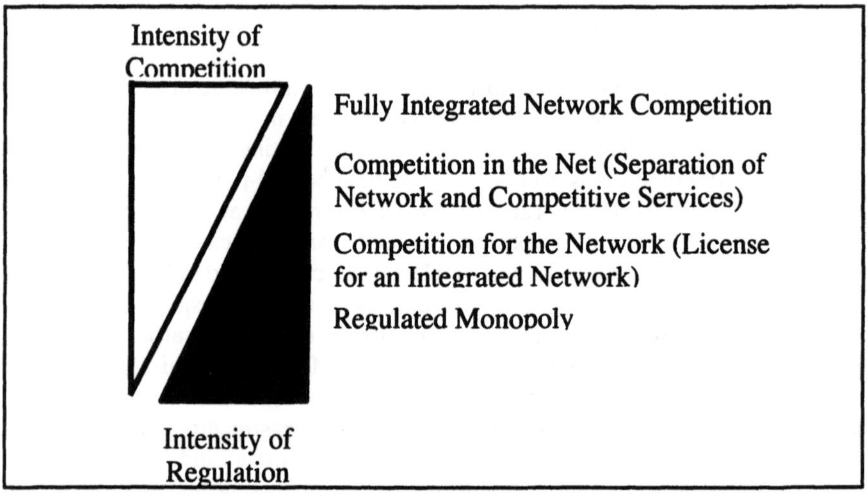

Source: WELFENS/GRAACK (1996), p. 133.

Fig. C9: Layers of the Infrastructure Business Relevant for Competition

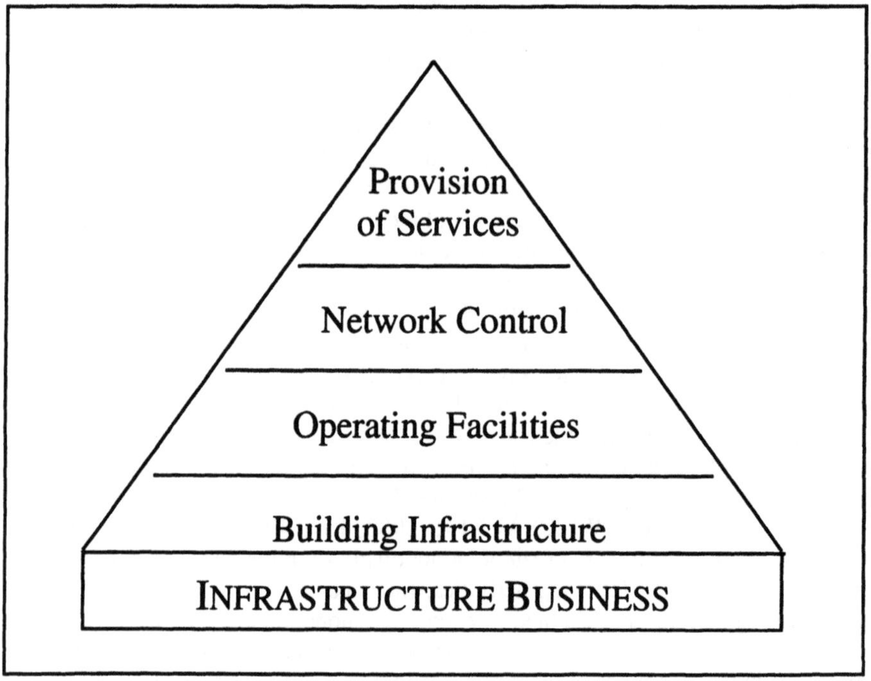

Another problem concerns market power derived from exclusive control over essential facilities associated with high sunk costs, allowing the incumbent to fend off newcomers using strategic pricing. Problems for newcomers also exist if the quality of service is impaired by feasible bypass options. A convenient way of preventing market entry is "vicious vertical integration", which is to combine infrastructure operation with an exclusive control system that typically leads to high sunk-costs and therefore to a natural monopoly. Independently integrated European railway and telecommunication control firms - each covering several countries, thus exploiting scale economies - could stimulate competition in both sectors. To illustrate this point, we make the hypothesis that in each European country, airline traffic control were vertically-integrated with the nationally dominant airline. The number of effectively competing airlines at given time intervals would drastically drop in the long-run. This points out the importance of the neglected long-run problems of vertical integration in infrastructure.

The network theory of infrastructure has mainly emphasized two points which complicates the achievement of efficient allocation via the market:
(i) positive network externalities on the demand side, which we will study in a first instance, (ii) capital intensive infrastructure networks often exhibit economies of scale: therefore, falling average costs imply both a deficit problem at the socially optimal quantity and a problem of ruinous competition. Government ownership and regulated national or regional monopolies was the institutional answer to this problem (with the exception of telephony in those countries where individual rental rates combined with capacity pricing avoided - via a two part tariff - the deficit problem). This, however, was meant to ignore problems identified by public choice, namely the self-interest of governmental bureaucrats in the overmanning of public firms and in the impairment of market access for newcomers. The major problem facing private firms, clearly, was even more difficult for state-owned firms where the ultimate owner is a dissipated group of poorly informed taxpayers. The political voting mechanism can be no effective substitute for a functional capital market. Besides, national ownership reduces options for the internationalization of the telecommunication market, a market where R&D is relatively intensive making it an appropriate candidate for foreign direct investment (as markets for technology are very imperfect).

From a theoretical point of view, falling costs are no justification for the presence of a regulated monopoly. A regulated monopoly for activity A is only appropriate if there are high sunk costs in this activity, (e.g the local fixed network or the rail track when dealing with railway companies) and if there are no "superior" by-pass technologies available by newcomers for the respective activity. High sunk costs coupled with falling average costs create the problem of an incontestable market in infrastructure. When faced with potential market-entries, the incumbent operator could disregard all sunk costs which cannot be recovered upon leaving the business. Hence, in a strategic pricing battle, high sunk costs could lead to a limit-pricing strategy fending off newcomers even though they were more efficient than the incumbent operator himself. If firm A undertakes simultaneously activity A with other activities, say B and C, it becomes an integrated monopoly assuming the presence of those regulations which act as barriers to market-entry within the respective activities; in addition to activity A, activities B and C become monopolized. In the past, this problem was greater in the telecommunication sector, where only network operations represent high sunk costs (powerful mobile telephony was not available before the early 80s).

Because sunk costs occur both in hardware investment and in the know-how of its users, an inferior technological trajectory actually becomes an optimal one as so many users invested in complementary skills (DAVID and BUNN, 1988; BENHABIB and SPIEGEL, 1992). Similarly a sub-optimal institutional setting could be made optimal if path-dependency is ruling. This means that many firms and individuals with high sunk costs invest into complementary "assets".

4.2 Domestic Competition and Foreign Direct Investment in Telecommunications

Competition is increasingly characterizing OECD countries in telephony. This holds for (A) network operation in a) local, b) national and c) international voice telephony, B) data transmission and C) value-added services. After 1998 almost all EU countries will have privatized telecoms firms and competition in telephony will be normal. Due to technological progress the fixed network no longer is the only access to customers (see Fig. C10). Wireless DECT technology could be offered by several firms in order to get access in the local loop. Satellite technology is another possible direct access to the customer. In addition to the fixed network of the dominant telecoms operator there are also alternative networks - used so far only internally - of gas, railway, energy, water and highway companies which could enter the public switched telephony market with some additional investment. Cable TV companies also have networks which can route telephone traffic in addition to other signals. Thus there could be several telecoms firms in industrialized countries which offer cheap and reliable telecoms services to households and industry. Ubiquitous and cheap telecoms networks will facilitate telelearning, telebanking and communication.

As long as telecoms operators are both in the network business and in the services business there will be room for cross-subsidization, namely that extra profits in one activity (e.g. network business) are used to artificially reduce prices in other fields (e.g. telecoms services). Cross subsidization is a double distortion of competition because one activity is overpriced relative to costs, while another activity is subsidized. This, however, is not in accordance with efficiency criteria which require that relative prices reflect relative marginal costs.

Competition means equal prices for equal products or services. Given the large international price discrepancies in international calls, it is clear that there is rather limited competition in this field. Traditionally high profits in international telephony will be no longer possible for the formerly dominant network operator after liberalization and free market entry has been implemented; call back services which exploit international price differentials for international calls (e.g. US-EU cheaper than EU-US calls) created pressure for dominant operators in Europe to lower prices (GRAACK, 1996, 1997).

International calls will be routed via those countries which have low international call prices and transit charges (access charges and conveyance fees). Countries with effective liberalization - such as the UK and Sweden - are particularly to benefit from this.

Fig. C10: Alternative Telecoms Operators

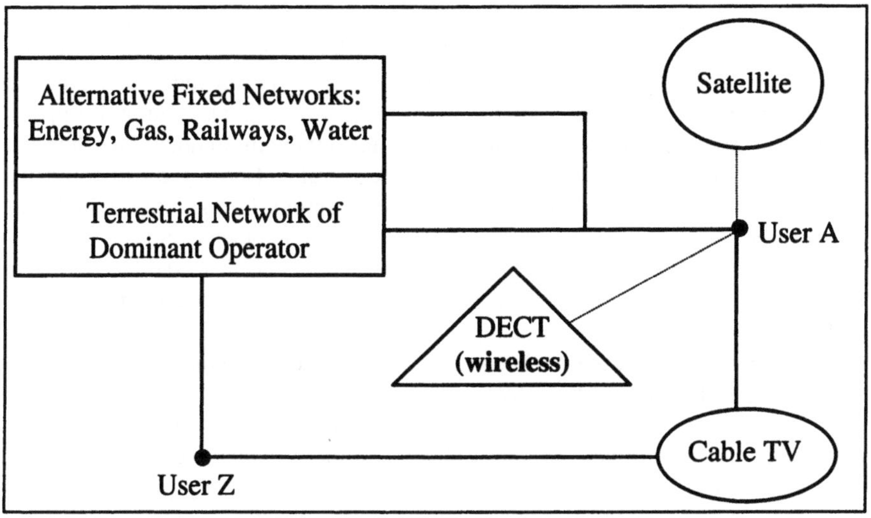

5. Regulatory Issues

5.1 Regulatory Framework and Regulatory Policy

Regulatory policy - possibly asymmetric in favor of newcomers - mainly concerns the following areas:

- Numbering plans ensuring number portability so that switching from one telecoms company to another is not impaired. In France the regulator (independent as of 1997) proposed a new scheme in 1996 where the telecoms user can determine the long-distance company by the first digit so that access to such an operator is possible on a call-by-call basis.
- Pricing rules which aim at promoting cost-oriented pricing and technological progress (e.g. via price caps).
- Interconnection rules which primarily refer both to technical protocols and pricing for network access.

- Universal service obligations which have to be defined and to be financed (e.g. from tax receipts, on the basis of market shares or from access charges).
- Frequency allocation in the case of mobile telephony.
- Technical regulations which allow interoperability of networks.
- Rules for consumer protection.

Regulated industries are subject to the problem of regulatory capture as emphasized by the Chicago School (POSNER, 1974; STIGLER, 1975): The industry's representatives influence legislation and regulation in a way which effectively impairs market entry and thus allows rent-seeking by the regulatee. British-style regulation adopted under the Thatcher government was a new attempt in Europe to implement regulation as a means to achieve and maintain competition when the initial situation is characterized by a dominant operator. Hence the British Office of Telecommunication (OFTEL) implemented asymmetric regulation which imposed more constraints on the dominant operator than on newcomers.

Regulatory bodies should be politically independent so that credible rules can be enforced. Similar principles of regulatory policy will be adopted in all EU countries. Common principles of regulation might be even discussed within the OECD, which has increasingly become interested in the telecommunications sector (YPSILANTI and KELLY, 1993).

EU regulation provides a framework within which national regulators can operate in accordance with national and EU law and the principles of the single market. Competition in the presence of a dominant operator is only possible if there are clear interconnection rules and if open network provisions (ONP) are implemented. Since the 1980s the European Commission has emphasized the principle of ONP. In 1996 the EU adopted a common position (European Commission and European Parliament) that licenses for telecoms operators cannot be refused by national authorities except in four cases: (1) in order to give a licensee access to a radio frequency for mobile telephony or if there is a temporary shortage of numbers, (2) to give the licensee access to private or public land, (3) to enforce universal service obligations, (4) to ensure interconnection in the case of a dominant fixed network operator. A major regulation authority at the EU level is the ONP Committee which operates under the chairmanship of a Commission member and is composed of representatives from all member countries. Decisions are taken by qualified majority voting but can be vetoed by the Commission, which in turn can be overruled by the Council of Ministers, otherwise the Commission's veto stands. The ONP Committee wants to ensure interoperability and hence has some arbiter function in the field of ONP leased lines directive. By 2000 Competition in all EU countries had increased, however in local telephony competition is still weak in most continental EU Countries.

Appendix I: Monopoly Versus Competition in International Telephony

Competition in international telecommunications will bring low prices and hence higher quantities (q in the home country, q* abroad) of telephone calls than under monopoly. In the following figure the market for international telephone calls is analyzed, where analysis clearly shows that competition brings economic advantages. It is assumed for simplicity that both countries have identical demand curves (DD and DD*; * for foreign variables) and that marginal costs k' at home and k'* abroad are identical; we disregard fixed costs. The Cournot solution is determined by the intersection of the marginal revenue curve R' with the marginal costs curve k' (R'* and k*' abroad), so that quantity is q_o (q_o* abroad) and the corresponding monopoly price p_o (p_o* abroad). The home country will have q_o outgoing calls and q_o* incoming international calls, where one may assume that each call [of firms] represents an exchange of valuable information that in turn contributes to value-added: directly as an immaterial input in the production process or indirectly by stimulating trade. Revenue in telecommunications is $p_o q_o$ at home and $p_o^* q_o^*$ abroad (Fig. C11), where we disregard international settlement conditions which basically give the country receiving calls a share of the revenues generated in the country with the outgoing calls.

If both countries switch to competition, the price at home and abroad is lowered to p_1 and p_1*, respectively, while quantity is increased to q_1 and q^*_1: The international exchange of information - as part of invisible trade - is raised. The number of calls and exchange of information, respectively, has strongly increased: instead of the distance BB* it now is equivalent to the distance EE*. For society the benefit from telephony is directly proportional to the amount of information exchanged.

It may well be - as is the case in the figure shown - that the revenue under competition ($p_1 q_1$) is smaller than in the monopoly case ($p_o q_o$). This will create resistance of the telecoms operator against a switch from monopoly to competition. However, in the long run, given the existence of call-back services and likely changes in international settlements procedures, competition is inevitable anyway.

Fig. C11: Monopoly vs. Competition in International Telephony: The Static View

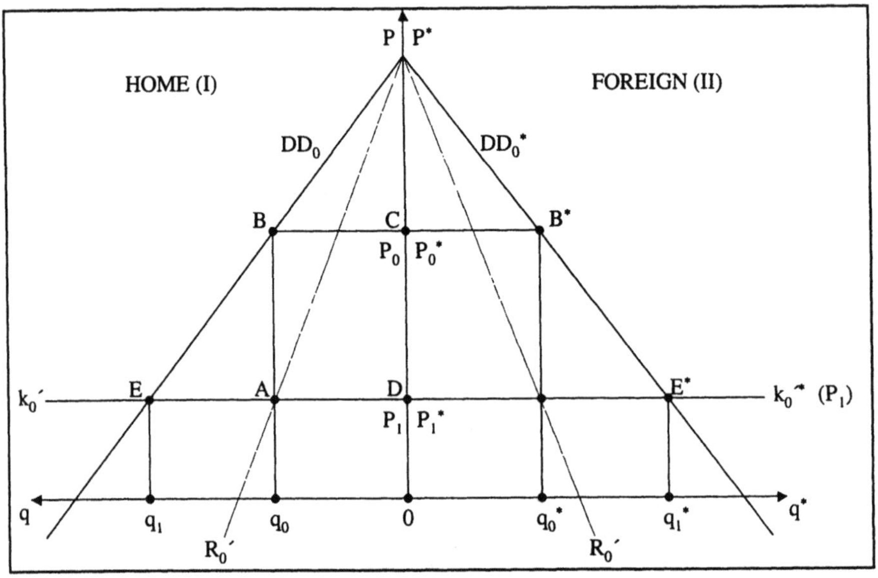

If only one country is liberalizing one may assume that calls increasingly will emanate from the country with the cheaper calling price. For example, a multinational company with a subsidiary abroad could decide that regular telephone discussions should always be organized in a way that the side with the lower telephone costs is the one paying for the call. This effectively means that in the country (with competition) with the lower price the demand for international telephony would shift outwards (DD^*_1 instead of DD^*_0), while in the other country there would be a corresponding inward shift of the demand curve (DD_1 which for ease of exposition is identical with the R'_0 curve). In the monopoly country the telephone price would remain at P_1 ($=P_0$) and the revenue would fall. By contrast the country with competition would have a new quantity q_1* of outgoing calls and generate additional revenue equivalent to the area A*B*I*H* (Fig. C12); international settlement procedures would give part of this increase to the monopoly country. So, in a world in which partner countries are liberalizing their telephony it obviously is not very sensible to maintain monopoly prices.

Fig. C12: Monopoly vs. Competition in International Telephony: The Dynamic View

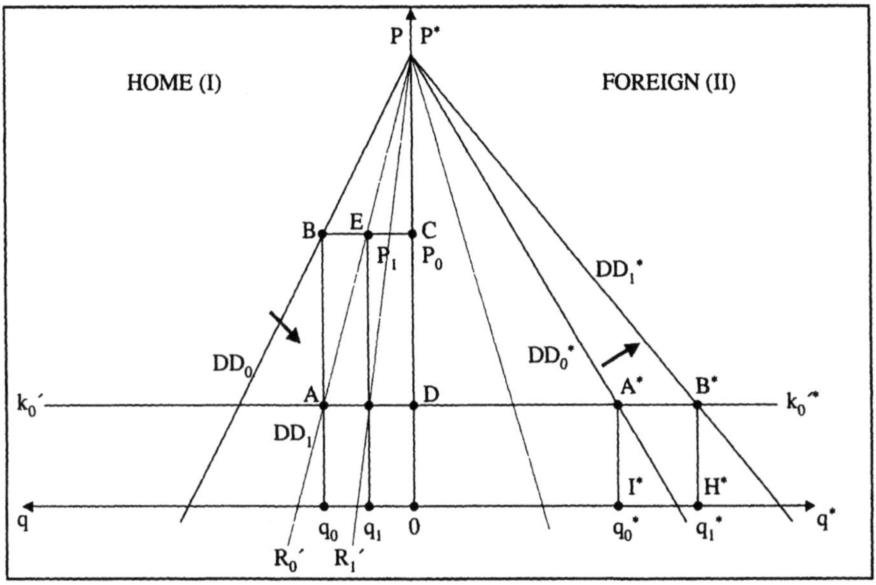

Appendix II: Deregulation Dynamics

Fig. C13: Deregulation Dynamics

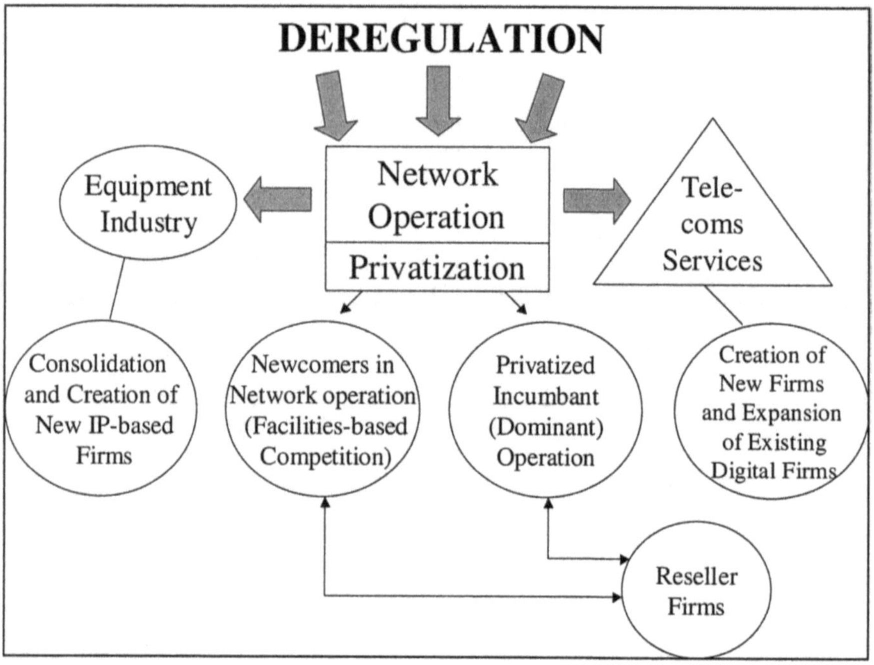

Appendix III: External Effects of Telecoms Network Expansion

Positive external effects of information would imply that the marginal social utility of information/the telecoms network is higher than the level expressed by market demand DD_0. Adding the positive external effects to the demand schedule DD_0, we obtain the demand schedule DD_1 relevant for optimum allocation. Assume for simplicity's sake that DD_1 intersects the abscissa at twice the value of the intersection point of the DD_0 curve, such that DD_0 coincides with the marginal revenue curve relevant for DD_1. Without external effects on the demand side, optimum output is q_1; taking into consideration the full social utility of information leads to an optimum output of q_2, which is clearly much higher than q_1 and very much higher than the monopoly solution q_0. The extra benefit for society in terms of consumer rent is the equivalent of the area ZHJ minus the triangle HGE_2.Optimal allocation could require state subsidies for the operator of the network(s). The subsidy per unit of output should be equivalent to GE_2 in this case, where we assume that adequate price regulation or effective potential competition could induce the firm to supply q_2 if a corresponding government subsidy is guaranteed. Moreover, there will be positive network externalities in the early stage of network expansion. If the network operator is moving on the marginal cost curve from K to E_1 , the existence of positive network externalities means that the initial demand curve DD_0 is shifting outwards (DD_1) since positive network externalities exist on the demand side; with a rising number of line subscribers the marginal utility of communication will increase. In a monopoly the interaction of falling marginal costs and network effects will bring about a lower price P_1 (instead of P_0) and a rising output q_1.

Fig. C14: Network Effects in Telecommunications (endogenous shift of DD_0 curve; here marginal revenue R'_1 of DD_1 is $=DD_0$)

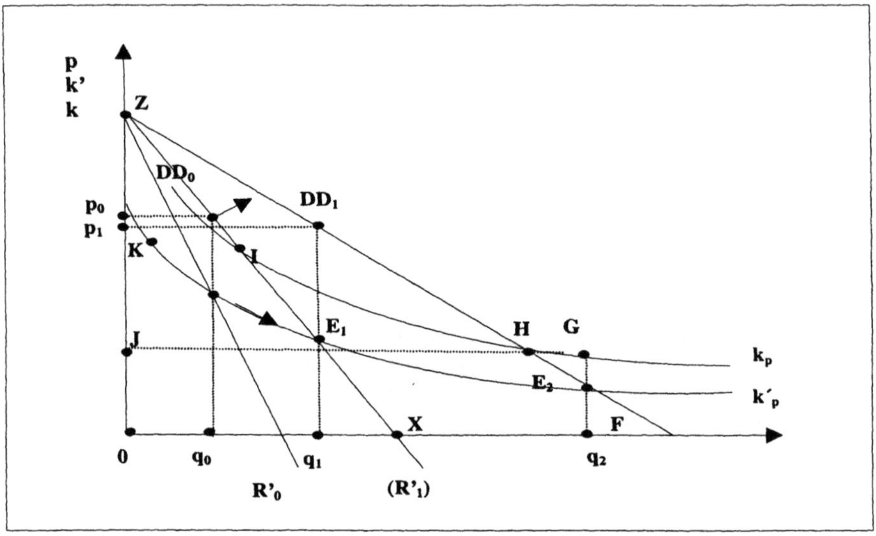

Source: WELFENS (1997), p. 111

Appendix IV: Transfer of Market Power in Interdependent Markets

Fig. C15: Transfer of Market Power in Interdependent Markets

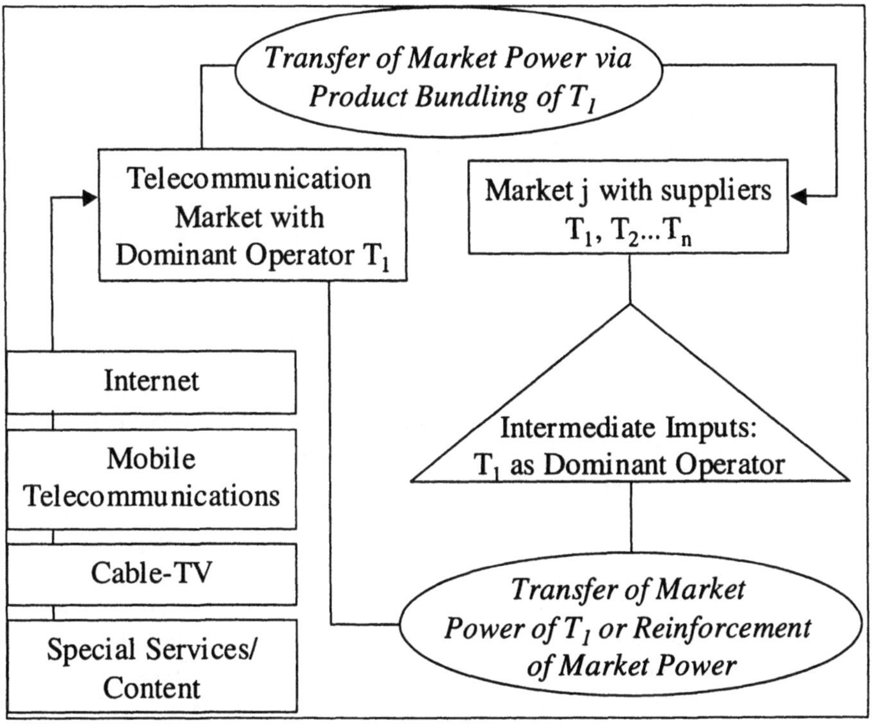

Appendix V: Telecommunications and Trade: Theory, Gravity Equation Analysis and Implications for Economic Policy[1]

International trade among OECD countries and worldwide has strongly increased after World War II. Explaining trade traditionally rests upon differences in factors supplies and relative prices on the one hand (survey CAVES/FRANKEL/JONES, 1990), on the other hand Schumpeterian influences associated with product cycle trade or trade with differentiated products plays a major role; the latter has stimulated models with monopolistic competition (eg DIXIT/NORMAN, 1980; KRUGMAN, 1979; LANCASTER, 1980; HELPMAN/KRUGMAN, 1990). Technological aspects also are important for goods characterized by static and dynamic scale effects. A very important link exists between per capita income and trade: on the one hand rising per capita income will raise the demand for differentiated products, on the other hand an increase national output will raise the demand for intermediate and final products. Thus the impact of a rising aggregate output can – with population given – reflect two different links between trade and income.

From a theoretical point transaction costs and information costs play an important role for international trade but little empirical research has been made about this. Transaction and information costs basically are like tariffs so that exports and imports are affected negatively. At the same time such costs add to overall costs which reduces overall profitable output.

Establishing and expanding international business always involves the use, storage and processing of relevant information – information about suppliers, markets, prices, technology trends and export or import markets. Precise and adequate information is required for choosing optimum market penetration strategies. With reduced international communication costs expanding into international markets becomes more easy, and therefore modern and more efficient telephony systems can be expected to have a positive impact on trade. From a theoretical point – taking into account that foreign direct investment can be an alternative for serving foreign markets which depends on the size of firm internal transaction costs relative to the costs of market transactions – technological progress in telecommunications might create a bias in favor of more firm internal transactions. Indeed, in a dynamic pespective there might be both more foreign direct investment and more trade, the latter reflecting creation of a larger market radius as a consequence of falling international information and transaction costs while a rise of FDI could reflect the interplay of the enlargement of market radius and of reduced firm-internal transaction costs.

Given the gradual EU liberalization of telecommunications services – beginning in 1984 already and culminating in the 1998 deadline for the liberalization of network operation and telecoms services – one may anticipate that international information and communication costs will fall; it would indeed be interesting to assess the potential impact of telecommunication systems on trade in Europe and

---

[1] This analysis draws on joint work with Andre Jungmittag.

elsewhere. In this paper we want to focus on the link between the telecommunication system and trade within the gravity approach.

EU integration has created a customs union by 1968 and thereafter the single market until end-1992 so that trade barriers have been reduced over time. Among the important elements of the single market program the opening-up of public procurement is rather important; moreover, the elimination of customs controls has reduced international transaction costs. This could mean that the role of information costs for trade has increased over time.

International trade relations can be modeled in various ways. Particularly prominent is the gravity equation which is based on market size in the importing and the exporting country on the one hand, on the other hand distance plays a crucial role. Certainly, there are also other important variables, including the role of telecommunications which will be analyzed here.

The natural point of departure is that international telephone calls are an important element for finding out about and arranging sales abroad or profitable imports. The number of telecommunication minutes is one potential proxy variable for measuring international telecommunications. From an economic perspective increasing international telecommunication links amount to a reduction of information and trading costs. While it is true that the internet has increasingly become an important source of information at the beginning of the 21st century it is realistic to assume that traditional telephony has been the dominant source of international information in the 1990s – and international telephony indeed will continue to play an important role in the future. From a data perspective it is important in principle that data on international telecommunications are available in a distinct way: we know how man international telephony calls went from country i to j (there is, however, depending on national competition conditions and prices, respectively, a potential bias with respect to originating calls coming relatively more often form i or j). With internet data traffic the direction of information diffusion is more difficult to assess. This holds particularly since preferred routings often use the US; even intra-Asian internet traffic is partly routed via the US which offers a cheap hub function due to low prices in leased lines and IP services (FCC, 2000).

By taking into account the role of telecommunications in a gravity model – as a new element in research – we can offer a better explanation of international trade than in previous approaches. Moreover, we also have the opportunity to come up, based on refined empirical analysis, with more adequate forecast analysis which could be particularly relevant for eastern Europe and other areas in the world economy. Assuming that telecom density and the use of telecommunications are proportionate we can furthermore provide an estimate of the trade potential for the case that telecom densities should increase in the future in eastern Europe and Asian (and other) NICs. With respect to eastern Europe EU accession can be expected to stimulate the growth of telecommunication penetration rates.

Section 2 looks into the trade related aspects of telecom liberalization and of the internationalization and modernization of telecommunications. In section 3 a simple augmented trade gravity model is presented – taking into account additionally telecommunications; and we discuss some basic methodological issues and problems related to data. In section 4 we present some empirical findings while the final section gives the core implications of our analysis.

The main results of our analysis is that we can provide for the first time a significant impact of telecommunications on trade. An important finding is that the GDP elasticities are lower than in previous studies. From this perspective the EU should expect less positive impact of GDP growth in EU-15 and eastern Europe on accession countries than is suggested by traditional gravity equation studies – unless the accession countries make considerable efforts in modernizing the telecommunications sector. This points to the crucial need of EU-15 to encourage accession countries to raise telecom penetration rates. Similarily, an efficient expansion of the telecommunication sector in the Balkan States might have a higher priority in the future; while the Balkans Stability Pact emphasizes the growth of regional trade (WELFENS, 2001) the twin role of telecomms modernization has not be recognized, so far mainly the infrastructure aspect has been emphasized. To the extent that increasing trade intensities stimulate economic growth it is obvious that expansion and modernization of telecommunications network indeed is a double stimulus for economic growth.

Theoretical Aspects of Telecommunications and Foreign Trade

From an analytical perspective with a focus on the internationalization of the economy it is crucial to understand the significance of information costs and the role of telecommunications, respectively. Cheaper and faster communication can stimulate market transactions and also widen the radius of international trade. From this perspective the enormous fall in national and international telecommunication costs after the 1998 liberalization of telecommunications could have important long term impacts. In a very basic sense international telecommunications is a fundamental basis of information acquiring of firms in the tradables sector and hence the use of international telecommunication links is an indicator for information gathering in international goods markets. We assume grosso modo that the intensity of international telecommunication between country i and j is an indicator for the exchange of information leading to trade. While bilateral telecommunication traffic might in some cases be biased by immigrants from the "partner country", at a rather general level we can assume that this should not create a serious bias in the data. One might have also the caveat that private international telecommunications is not very relevant for trade, but we indeed can assume than even in private international telephony often trade-relevant information is exchanged or disseminated; eg information about product innovations, about new fashion trends, about differences in international prices or about travelling opportunities. It is, however, obvious that telephone calls of firms are more relevant for international trade than residential calls. Unfortunately, the available statistics do not allow to distinguish between international traffic of private households versus firms.

As regards the direction of traffic one may assume that strong outgoing traffic from i to j (i-j-traffic) is more important for i´s exports to j than telephone traffic going from j to i. Whether this hypothesis is reasonable can be tested. An alternative is the sum of telecommunication contacts i-j and j-i which then would be a proxy for "economic information distance".

A potentially serious data bias might come from differences in market conditions at home (country i) and abroad (j). Naturally, a country with a monopolized telecommunications sector will have relatively little outgoing international traffic – as prices for international telecommunications are high – and relatively much inbound international traffic. From this perspective it is not surprising that the US, having a rather competitive telecommunications sector since 1984 and recording a gradually increasing deficit in the services current account strongly demanded that other countries should open up their respective markets (SPINDLER, 1999).

As regards market opening in telecommunications there are various approach in the EU where the EU has set the key date of 1998 (WELFENS/GRAACK, 1996; WELFENS/YARROW, 1996; PELZEL, 2000). The EU´s liberalization in the field of network operation and voice telephony – with grace periods for Luxembourg, Ireland, Greece and Portugal – has created intensified competition in every country; with considerable variations as regards the major impulses, eg stemming from newcomers from the energy sector, cable TV operators and wireless technologies. Long distance costs and international telecommunications costs have fallen strongly as there were parallel quasi-reciprocal market entries when national markets were opening up: As every post-monopolist national operator was eager to generate additional revenues in a period in which domestic markets were opened it was only natural to expand internationally. Due to enormous price reductions in international telecommunications in 1998/99 there was a strong increase in international telecommunications volume. As regards international telecommunications in Western Europe there are still some bottlenecks, however, increasing digitization and growing transmission capacities, built partly by newcomers, should remove these bottlenecks in the medium term.

As regards eastern Europe the situation is partly special in countries which have still majority ownership of the telecommunications operator. To the extent that government still is the main owner of the monopoly operator government can determine directly the pace of upgrading of the telecommunications infrastructure in the transition stage. Alternatively, government can set framework conditions for private newcomers which encourage high investment in telecommunication infrastructure; here, regulatory policy – including determination of price-caps – clearly is important for investment decisions. There could be temporary trade-offs for competition and network expansion to the extent that rather high prices, in an environment of limited competition, might enhance investment in international telecommunications. At the same time one must take into account that monopoly prices in international telecommunications amount in effect to taxing international communications. For transition countries opening-up the modernization of telecommunications and the rebalancing of tariffs – leading to cost-based pricing – is important not only for international arbitrage but more generally for facilitating access to international market information as well as a basis for establishing international production networks within multinational companies.

Gravity models became popular since the analysis of LINNEMANN (1966) whose analytical point of departure was rather economic geography than standard economic trade theory; the gravity approach has found other researchers who contributed to the theoretical foundation of what was first mainly a good empirical fit in analyzing international trade relations not only on the basis of GDP in the exporting and the importing country - or per capita GDP in the exporting country and the importing country, respectively - but also taking into account distance (BERGSTRAND, 1985; ROHWEDER, 1989; DEARDORFF, 1998; EVENETT/KELLER, 1998; FEENSTRA/MARKUSEN/ROSE, 1998). As regards GDP in the exporting country (Y_i) one may assume that the higher real GDP the better are the opportunities to exploit economies of scale; and with population given one also could argue that a rise of real per-capita GDP will bring about more product differentiation and hence more product varieties which can be exported. As regards foreign GDP(Y_j), that is GDP in the importing country it is clear that a rise in real foreign GDP reflects a rise in aggregate demand which naturally will raise imports. As regards the distance variabl (D)e: From an economic perspective distance is not so much meant to measure physical distance in terms of km but transportation costs – and risks- in space. The gravity model can be specified in its basic version in the following way if one is to focus on exports X_{ij} from country i to country j:

$$\text{(I)} \quad X_{ij} = \beta_1 Y_i^{\beta 2} Y_j^{\beta 3} D_{ij}^{\beta 4} u_{ij}$$

Equation (I) can be rewritten in logarithmic form where u_{ij} is a normally distributed error term:

$$\text{(II)} \quad \ln X_{ij} = \ln \beta_1 + \beta_2 \ln Y_i + \beta_3 \ln Y_j + \beta_4 \ln D_{ij} + \ln u_{ij}$$

There are many applications to and variations of the gravity model in the literature, e.g. including a focus on trade or foreign direct investment inflows. (DEUTSCHE BUNDESBANK, 1999) and patent indicators for approximating technological performance (JUNGMITTAG, 2001). FREUND/WEINHOLD (2000) found positive empirical evidence on the impact of internet host density on trade.

A modified approach presented in WELFENS/JUNGMITTAG (2001) is obtained when one includes a variable for international telecommunications – additionally there are two dummy variables (β_5 and β_6) for common language and EU membership. The hypothesis is that the more intensive international telecommunications are the more trade-relevant information is flowing across borders. This could concern both information relevant for international arbitrage, information about product innovations and information about product varieties or cheaper or new intermediate products. We clearly expect a positive coefficient for ß7, the coefficient for the international telecommunications variable.

As a proxy for international telecommunication contacts between i and j the authors used $\ln TM_{ij} + \ln TM_{ji}$; so there are two new coefficients β_7 and β_8; in a second specification the authors use $\ln (TM_{ij} * TM_{ji})$. The reason to include a variable which reflects both communications going from i to j and that from j to i is not only related to economic reflections, namely that for trade between i and j telecommunication minutes going from i to j are potentially as important as outgoing calls from j to i. The combined use of ij-minutes and ji-minutes also is necessary to adequately cope with the bias problem that emerges in the context of large differences in prices for international telephony. Typically, the US – having had competition already since the 1980s – will have low prices in international telecommunications which implies that as long as (continental) EU countries still have a monopoly structure, namely until 1998, there is a much larger probability of telecommunication calls outgoing from the US to EU countries than from EU countries to the US. The authors use data for the period 1995, 1996 and 1997 for 12 OECD countries (having the full set of necessary bilateral trade data); bilateral telephone minutes come from the International Telecommunications Union and bilateral trade figures are from the IMF's Directions of Trade Statistics.

Tab. C1: Estimation Results for the Simple Gravity Model

	(1) 1995	(2) 1996	(3) 1997	(4) 1995	(5) 1996	(6) 1997
Constant	2.894	3.039	3.227	3.673	3.831	3.893
	(5.865)[1)]	(6.015)	(6.130)	(7.999)	(8.387)	(8.210)
$\ln(GDP_i)$	0.942	0.957	0.956	0.607	0.606	0.600
	(21.222)	(20.631)	(19.545)	(7.743)	(7.763)	(7.288)
$\ln(GDP_j)$	0.802	0.798	0.798	0.491	0.444	0.435
	(18.669)	(18.073)	(17.516)	(7.252)	(6.559)	(6.009)
$\ln(DIST_{ij})$	-0.822	-0.851	-0.862	-0.599	-0.602	-0.601
	(-12.006)	(-12.358)	(-11.986)	(-7.024)	(-7.054)	(-6.776)
LANGUAGE	0.422	0.373	0.369	-0.040	-0.104	-0.110
	(2.601)	(2.190)	(2.240)	(-0.250)	(-0.683)	(-0.734)
EU	0.335	0.258	0.250	0.293	0.241	0.263
	(2.875)	(2.254)	(2.061)	(2.704)	(2.273)	(2.251)
$\ln(TM_{ij})$				0.295	0.219	0.214
				(2.110)	(1.582)	(1.772)
$\ln(TM_{ji})$				0.141	0.254	0.267
				(1.064)	(1.922)	(2.315)
Adj. R^2	0.888	0.879	0.866	0.907	0.904	0.892
F-test $\beta_7 = \beta_8$				0.348	0.018	0.059
Sign.-level				(0.556)	(0.894)	(0.809)

1) t-values in brackets. White's heteroskasticity consistent estimators of the variance matrix of the regression coefficients were used to calculate these t-statistics.

Distance is measured as air distance between capitals. GDP figures were taken from OECD National Accounts Statistics. Adjusted r squared is close to 0.88 and at least one of the two telephone variables always is significant at the 95% level. The estimations for 1995, 1996 and 1997, respectively, showed in each case an increase in r squared when the telecommunications variable was included. When ingoing minutes and outgoing minutes were combined in a multiplicative way the telecom variable became highly significant (>99% level) and the elasticities were around 0.23 – remarkably they increased over time. Endogeneity problems with the telecommunications variable were checked on the basis of a Hausmann test; it turned out that there was no endogeneity problem with the telecommunications variable so that the OLS estimators are useful indeed.

Implications and Options for Economic Policy

The analysis of the enlarged gravity model has shown that the income elasticities of GDP are lower than in traditional gravity models. Moreover, the international telecommunications has turned out to have a significant influence on European trade with goods. A rise of international telecommunications by 1% increases the volume of international trade by 0.2%. From this perspective it is the upgrading and expansion of telecommunications in all countries is crucial for improved exchange of information relevant for trade. As regards Europe the European Commission faces considerable challenges, given the fact that the Commission has been the main stimulus in opening-up of markets and deregulation. As regards eastern Europe and EU eastern enlargement, respectively, there is additional relevance of telecommunications policy. What holds with respect to eastern Europe is also relevant vis-à-vis the NICs and the developing countries: Here there is a major role for the World Bank.

The 1998 EU liberalization of telecommunications which brought enormous reductions in international telecommunication prices has trade-creating effects. A caveat are the high prices in mobile telephony prices which implies a disadvantage of the EU compared to the US: While it is true that US has the disadvantage that pricing policies to some extent discourage mobile telephony as both the outgoing caller and the incoming party have to pay – except for a few special tariff schemes and a minority of operators – the uniform low national prices in mobile telephony are a major advantage compared to Europe; it was only in 2001 that Vodafone and some other mobile network operators have offered subscriber options which include uniform prices in the EU; however, at a much higher level than the US. From the perspective of European integration it therefore would be desirable that national and supranational regulators encourage network integration on a European basis.

The US has benefited from enhanced competition in long distance and international telecommunications already since 1984. In Europe the UK was an early pioneer in liberalization, but market opening in the period 1984-90 was relatively modest since it was realized within a duopoly strategy. Given the broad 1998 EU telecoms liberalization one may anticipate a special stimulus for rising intra-EU trade at the beginning of the 21st century. As regards the impact of the introduction of the Euro the almost full coincidence in the timing of telecoms liberalization and the launching of the new currency should be carefully disentangled in future analysis of trade-creation in Europe.

Considering the rather surprising effect that US productivity growth in the 1980s and 1990s was significantly stimulated by the growth of international trade (MANN, 1998) while no significant link was observed for Germany might partly be related to the US lead in the liberalization of telecommunications. The hypothesis is that there is a double impact of telecoms liberalization, namely directly raising productivity and output as was shown in an empirical study for Germany by JUNGMITTAG/WELFENS (1996, 2001): Rising output and real income, respectively, should then in turn stimulate the use of telecommunications, and to the extent that international telecommunication traffic is expanding there will be a positive trade-creation effect raising output and real income, respectively.

Increasing research efforts with respect to the dynamics of telecommunications and the effects on productivity are desirable. A better understanding of both diffusion and innovation effects of the use of telecommunications and the internet could help to improve the efficiency of growth policy. Defining digital network analysis broadly thus means to analyze not just digital services and the growing use of computers.

As regards policy options government stimulus for the role of modernizing the telecommunications sector is important: Telecommunications policy is relevant for trade-creation and growth. The regulation of the telecommunications sector is quite important as the rules for the incumbant operator – eg price caps – on the one hand and on the other hand the regulations for network access and interconnection are highly important for the dynamics of competition in this sector. It should be pointed out that an important issue concerns the proper mix between a low regulation profile in the field of the internet and considerable regulation in the telecommunications sector. Inadequate regulations would impair investment and innovation in this sector. As regards the impact on trade competition in international telephony is crucial. The ITU – an international organization - has traditionally regulated international pricing as national monopoly operators had to cooperate to realize international calls (GRAACK, 1997). With market opening-up in all OECD countries and many newly industrializing countries the role of the ITU has become less important since ITU sets binding rules for governments, but leaves open private international interconnection agreements: If eg firms from country I, II and III, respectively, invade each other´s territory with investment in network capacity the initial bilateral monopoly situation is no longer relevant. Threat of foreign operators to invest more in capacity will have an effect on private interconnection agreements and hence on the splitting of revenues from international telephony. While international resale can generate some competition the more important element is facility-based competition. The fact that foreign direct investment has been increasing in the late 1990s lets us expect that prices in international telephony will continue to fall and this in turn will stimulate trade.

A potential application for our conclusions is the Balkans Stability Pact which aims at stabilizing countries in southastern Europe, not least by enhancing regional trade. Obviously a strong focus on the modernization of telecommunications and measures to stimulate competition in telecommunications in the Balkan countries could be an important element for enhancing regional trade. This basic insight also holds with respect to Asian NICs, Latin America and Africa.

High monopoly prices in telecommunications in many developing countries – deliberately imposed by government as a means to generate a sectoral trade surplus and high profits (accruing to government) – are quite doubtful if one takes into account the broader long term effects of such a policy: By impairing trade and a fortiori economic growth such a policy in effect will undermine overall revenue growth.

Finally, we may point out that our gravity model approach also could be applied to foreign direct investment. For multinational firms vertically integrated and for horizontally integrated firms international telecommunications is important. Telecommunications here is crucial for creating efficient international networks in sourcing, production and marketing. Here the hypothesis also is that there should be a positive significant impact of the telecommunications variable. It is clear that there are some difficult methodological issues with respect to empirical analysis, but one may hope that improving data sets and applying a variety of estimation procedures will help to clarify the issues in the long run.

D. Transatlantic Growth Differentials, ICT Dynamics, Fiscal Policy and Innovation Policy

1. Euro Weakness and Transatlantic Growth Differentials

The United States (US) recorded in the period 1991-2000 a formidable growth rate of about 3% p.a. where a considerable impulse for high and sustained growth stems from high investment in information and communication technology (ICT). Between 1993 and 2000 the US even recorded 4% growth. Roughly one-half of investment growth in the second half of the 1990s can be attributed to ICT which mainly is comprised of telecommunications and computers/PCs plus software. Euroland by contrast has grown only by about 2% in the period 1991-2000 where the large core countries of Germany, Italy and France have not even achieved 2%. Given the long and strong economic upswing in the US in the 1990s it was rather surprising that the inflation rate has remained very low in the US. Obviously, the non-accelerating inflation rate of unemployment has reduced in the US, but it is rather unclear why this should be the case.

The economic upswing in the euro zone in 1999/2000 was rather modest and already in 2001 the growth rate of Euroland fell below 2%. The EU has had lower investment growth than the US and also has failed to position itself adequately in the changing global innovation race (WELFENS ET AL., 1998; WELFENS ET AL., 1999); with the US no longer devoting almost 50% of its R&D budget to the military, the innovation race has accelerated and increasing specialization should have been realized in EU countries.

There is no doubt that the Maastricht convergence process has reduced economic growth since prior to the start of the euro and the ECB in 1999 many countries, including Italy, Germany, France and Spain had to reduce government-GDP ratios in order to reduce deficit-GDP ratios and subsequently excessive debt-GDP ratios. Germany, Spain and France were close to the 60% maximum for the debt-GDP requirement in 2000; among the large EU countries only Italy still had an excessive debt with 110.7% of GDP – but well below the peak ratio of 123.9% in 1994. While the Maastricht convergence process required fiscal retrenchment in some countries it also is true that downward interest rate convergence brought a reduction of the interest payments relative to GDP.

The euro has continuously lost value vis-à-vis the dollar since the start of the new currency in January 1999; the overall loss was close to one-fourth in the period 1999-2000. This is not so critical in the staggered introductory phase of the euro which will be completed with the introduction of coins and notes in early 2002. While the euro zone has witnessed a modest acceleration of growth in the late 1990s the US recorded a sustained upward shift in the expansion path of the production potential, while Japan recorded a downward kink in the growth rate of the production potential (COUNCIL OF ECONOMIC ADVISORS, 2001). Thus the question arises why there are such strong differences in the triad. Subsequently we will only focus on transatlantic differences where a major aspect concerns employment growth. Average annual employment growth reached 1.5 % in the US in 1991-2000, but only 0.5% p.a. in the euro zone (the same growth rate for EU-15 as well). Employment dynamics and labor markets, respectively, thus have played a major role for transatlantic growth differences. Another aspect concerns the investment-output ratio which fell in Euroland from a peak of 21.7% in 1991 to about 20% in all years from 1995-98 (20.8% and 21.4% in 1999 and 2000, respectively) while the US investment-GDP ratio increased from 16.3% in 1991 to 21.1% in 2000 – with a year on year increase in each year of the period from 1991-2000 (EUROPEAN COMMISSION, 2000).

The considerable depreciation of the euro raises the question whether this is a transitory development or a more long-term phenomenon. We will argue that it is likely to be a sustained problem unless policymakers in Euroland take adequate measures and revise their current policy stance.

The main effects of a strong real devaluation of the euro are the following:

- Stimulating Euroland´s exports towards the dollar area automatically makes Euroland more dependent on the US business cycle.
- The inflow of foreign direct investment from the US and other non-EU countries could increase since; following FROOT/STEIN (1991) foreign investors can – in a world of imperfect capital markets – acquire firms more cheaply than before.
- A risk premium emerges as reflected in the interest rate; until 2001 there was not yet a euro risk premium visible, but with a sustained devaluation of the euro such a premium might gradually emerge. A risk premium would raise the real interest rate and reduce the investment-GDP ratio.
- The inflation pressure in Euroland is increasing since the import of more expensive imported intermediate products and final products will translate into a rise of tradables prices.

The critical question indeed concerns the medium and long-term development of the euro. Based on transatlantic interest rate differentials and relative stock market prices WELFENS (2000) presented a robust out-of-sample forecast. In the following sections we want to shed further light on the devaluation issue.

Taking a closer look at the transatlantic growth differential we find several remarkable points in the 1990s (RÖGER, 2002; WELFENS, 2001):

- The USA has grown continuously faster than Euroland in the 1990s.
- High investment in ICT has considerably contributed to high US growth. While the share of ICT investment in national output has remained constant in the EU, the figure for the US has roughly doubled.
- The growth rate of labor productivity in high technology clearly outpaced that of the EU after 1993. Germany – representing one-third of Euroland´s GDP – faces not only a considerable gap vis-à-vis the US; worse yet is that the labor productivity in technology-intensive fields was lower than the average for the overall economy in the mid-1990s.

- The US has exploited the economic potential of the internet revolution much faster than the EU; both the user density (demand side) and the host density (supply side) have increased much faster in the US than in France, Italy and Germany. The US has a firm lead in computer density, being one-fifth ahead of Germany and one-fourth ahead of France; Italy has just half the US computer density and was matched in 2000 by the Republic of Korea.
- The fall in prices of ICT goods – relative to the GNP deflator – reached about 8% in 1981-94, but after 1995 it increased to 15% p.a which has considerably stimulated innovations, both process innovations and product innovations, including novel digital services. Falling relative prices clearly stimulate diffusion. This should benefit the US furthermore as the world´s leading computer producers and software firms are located in the US.
- A study by the OECD (2000) shows clearly that countries with a high R&D intensity in the ICT field also have a high R&D intensity for the overall economy. While Sweden, Finland, Korea, the US and Japan are leading economies from this perspective, Germany is only in a medium position. This OECD study, based on ICT dynamics with respect to employment, value-added, trade and R&D, suggests that Germany is among the lower third of the 29 countries.

Labor productivity growth and overall output growth has accelerated in the US in the second half of the 1990s as is shown in the following table. Germany's labor productivity growth reduced in the period 1995-2000 in comparison to 1977-95. The same is true for Japan. Only France and the UK achieved an improvement over time, and both countries did so on the basis of accelerated output growth (by contrast, in Germany output growth reduced)

Tab. D1: Labour Productivity Growth in Selected OECD Countries

	1995-2000*	1977-95
US	2,2	1,4
Japan	2,0	2,6
Germany	1,8	1,9
France	1,8	1,6
UK	1,5	1,9

Source: OECD (2000)

Tab. D2: GDP Growth in Selected OECD Countries

	1995-2000*	1977-95
US	4,00	3,00
Japan	1,25	3,50
Germany	1,75	2,25
France	2,50	2,25
UK	2,75	2,25

* estimate

Source: OECD (2000)

2. Theoretical Analysis

2.1 ICT Dynamics and Growth

It was not fully clear until 2001 whether high US growth in the 1990s was significantly related to high productivity growth in information and communication technologies. GORDON (1999, 2000) claimed that the acceleration of US productivity growth in the 1990s was mainly due to cyclical factors on the one hand, and to high productivity growth in the production of information technology goods (which would account for the remainder of accelerated productivity growth) on the other. A skeptical view also comes from KILEY (2000) who points to high adjustment costs associated with IT investment implying a reduction of productivity growth in a period of high IT investment. The contrasting view that both production of IT goods and use of IT contributed considerably to aggregate productivity growth in the 1990s has been particularly emphasized by JORGENSON/STIROH (2000), OLINER/SICHEL (2000), WHELAN (2000a,b), and the COUN-CIL OF ECONOMIC ADVISORS (2000, 2001), all of which provide some empirical evidence for their view.

With different researchers holding opposite views on an important empirical phenomenon, here the acceleration of productivity growth, the most straightforward way for clarifying the issue is theoretical research on the one hand, and more detailed empirical research on the other hand. As regards theoretical aspects we will subsequently present some reflections where we will argue that differences in productivity growth in the US and Euroland/Germany in the 1990s can partly be explained by differentials in ICT dynamics and the associated direct and indirect growth effects. As regards the empirical side it seems that three papers have brought clear evidence that both the production of IT goods and the use of IT were important for the acceleration of productivity growth in the 1990s. STIROH (2001) has presented an empirical study with industry-level data for the US. He shows that most IT-intensive sectors recorded significantly larger productivity gains than other industries. Using a battery of econometric tests he shows a strong correlation between IT capital accumulation and labor productivity where a novel decomposition of aggregate labor productivity is presented. The crucial conclusion is that virtually all of the aggregate productivity acceleration can be traced to the industries that either produce IT or use IT rather intensively with hardly any contribution coming from sectors that are less involved in the IT revolution. STIROH writes (2001, p.2-3):
"Industry-level data show that the recent U.S. acceleration in productivity is a broad-based phenomenon that reflects gains in a majority of industries through the late 1990s... For example, the mean productivity acceleration for 61 industries from 1987-95 to 1995-99 is 1.09 percentage points and the median is 0.67 percentage points. Nearly two-thirds of these industries show a productivity acceleration. Even when the particularly strong productivity industries that produce IT (or even durable goods manufacturing as a whole) are excluded, the data show a sig-

nificant acceleration in productivity for the remaining industries. This suggests that U.S. productivity revival is not narrowly based in only a few IT-producing industries... The productivity acceleration in the late 1990s for IT-intensive industries, for example, is about 1 percentage point larger than for other industries. Moreover, rapid IT capital deepening in the early 1990s is associated with faster productivity growth in the late 1990s, even after controlling for other input accumulation and productivity growth in the early 1990s. Production function estimates also show a significant and relatively large output elasticity of IT capital.... If cyclical forces were driving the productivity gains, one might expect these gains to be equal across industries or at least to be independent of IT-intensity... The data show a contribution to aggregate productivity in the 1990s from all three groups, although the vast majority comes from IT-related industries. For example, the 26 IT-using industries contributed 0.66 percentage point to the aggregate productivity acceleration and the two IT-producing industries 0.16. The 33 remaining industries contributed only 0.08. Once one accounts for reallocation of intermediate materials, the industries that either produce or use IT account for all of the aggregate productivity acceleration, with the other industries making a negative contribution to the acceleration of aggregate productivity growth in the late 1990s."

This finding is quite important since it points to a shift in Schumpeterian dynamics in the US and possibly in other OECD countries as well. In Europe only Sweden, Finland, Ireland, the UK, and the Netherlands show a high share of ICT dynamics, and it is unclear whether major EU countries such as Germany, France, and Italy can catch up. DAVERI (2001) shows in his empirical analysis for Europe that the UK, the Netherlands, Sweden, and Ireland are not much behind the US, but other EU countries have a considerable lag. Cross country differences in IT investment and accumulation rates are closely linked to growth effects from information technologies except for Ireland. CREPON/HECKEL (2001) also find for France that productivity growth is strongly associated with a small number of industries that make an intensive use of computers; all in all they estimate the contribution of computerization in France to have reached 0.7 percentage point in annual growth in the period 1987-1998.

According to DAVERI the growth contribution differs considerably across OECD countries (see subsequent table). DAVERI (2001) also finds that in terms of the share of IT capital (augmented by the software component), the EU faced considerable differences. In 1999 IT capital appropriated roughly 6% of overall value-added in Sweden and the UK, 5% in Ireland and the Netherlands, but only about 3% in Germany, Italy, France, and Spain, and in Portugal and Greece only 2.5%. In the US the share of IT capital in overall value-added was 8% (0.029 for hardware, 0.034 for software, and 0.016 for communications equipment) which is 1/5 of the value-added share for total capital. In Greece, Italy, Spain, and Ireland communications equipment absorbed about 2/5 of the IT capital share of value-added. In most other EU countries the distribution of value added to the various IT categories was more similar to that found in the US.

Tab. D3: The Growth Contribution of IT Capital and Its Components

	(1) 1991-99 IT	(2) 1991-95 IT	(3) 1996-99 IT	(4) 1991-99 HW	(5) 1991-99 SW	(6) 1991-99 TLC
USA	0.94	0.53	1.45	0.50	0.36	0.08
Ireland	0.64	0.38	0.96	0.30	0.12	0.22
Denmark	0.52	0.42	0.65	0.29	0.14	0.09
Netherlands	0.68	0.65	0.72	0.33	0.22	0.13
UK	0.76	0.43	1.17	0.39	0.26	0.11
Portugal	0.43	0.39	0.49	0.18	0.05	0.19
Austria	0.45	0.47	0.43	0.23	0.12	0.11
Spain	0.36	0.38	0.34	0.17	0.06	0.14
Greece	0.34	0.25	0.46	0.12	0.04	0.18
Finland	0.45	0.21	0.74	0.27	0.10	0.08
Belgium	0.48	0.48	0.49	0.23	0.14	0.11
Sweden	0.59	0.38	0.85	0.38	0.13	0.08
Germany*	0.49	0.54	0.45	0.24	0.12	0.13
France	0.41	0.40	0.44	0.20	0.11	0.11
Italy	0.31	0.28	0.35	0.15	0.05	0.11

Notes: IT = HW+SW+TLC = Hardware+Software+Communications equipment. Data in percentage points.
* Germany = 1992-1999
Sources: Daveri, F. (2001): Information Technology and Growth in Europe, p. 6.

While the US recorded an acceleration in growth contribution from IT capital, namely from 0.53 in 1991-95 to 1.45 in 1996-99, Germany had a rather flat contribution from IT capital, namely 0.5. Insufficient structural change, the monopoly of telecommunications until 1998, and lack of cheap internet rates (including the refusal of Deutsche Telekom AG (except for the second half of 2000) to offer a flat rate for standard telecommunications users) could be main elements in explaining the modest IT contribution in Germany (WELFENS, 2001a, 2001b). Comparing the US and the EU it is quite obvious that computer density in the Community is much lower than in the US. Germany reached ¾ of the US computer density, France was slightly weaker, and Italy, with about ½ of US computer density, was just at the level of Korea in 1999. In the EU only the Scandinavian countries had a computer density comparable to the US. It is noteworthy that Sweden's tax policy contributed actively to raising computer density by giving firms tax incentives to sell used PCs at discount prices to employees.

In the digital economy the inventory-output ratio is lower than in the traditional economy since computerization allows improved production planning and logistics. At the same time the digital economy consists to a considerable extent of digital services where the supply elasticity is obviously very high. In the case of services provided via the internet (e.g. software (application sharing services) or music) the supply elasticity is extremely high. This could contribute not only to higher growth in fields where demand follows a logistical expansion path over time, it could also reduce the inflationary pressure for any given money supply growth. That is in economic upswings the inflation rate will increase less than in the traditional economy. This seems to indeed have been the case in the 1990s. CREPON/HECKEL (2001) find that computer use and total factor productivity gains in the ICT sector have reduced the inflation rate by 0.3 and 0.4 percentage points in the period 1987-1998, a considerable impact with respect to the average inflation rate of 1.4% in this period in France.

JORGENSEN/STIROH (2000) have argued that US growth resurgence is partly related to an increasing use and production of ICT. OLINER/SICHEL (2000) have also concluded that the revival in US productivity growth is strongly related to information technology dynamics. About 2/3 of the rise in US labor productivity in 1996-99 can be explained by an increasing use and production of information technology. These 2/3 can in turn be partly assigned to capital deepening and partly to higher total factor productivity growth. DAVERI (2001) argues that growth in EU countries is also partly an information technology story. The previous table shows that IT capital accumulation can explain a considerable part of cross-country growth gaps. The shares of the growth gaps explained by IT capital is roughly 25-30% of the total for six EU countries (Germany, France, Italy, UK, Sweden, and Belgium), but this fraction is larger for Denmark (90%), and Greece, Spain, Portugal, and the Netherlands, and Austria and Finland (50-60%). As regards capital input, differences in the overall contributions of capital cannot explain much of the EU growth gap vis-à-vis the US. It is noteworthy that the growth gap observed in Italy, Germany, France, and Sweden vis-à-vis the US is largely explained by gaps in the contribution of labor. Following similar experiences in the 1960s, 1970s, and 1980s (DOUGHERTY/JORGENSEN, 1996), the growth contribution of labor was negative in many EU countries in the 1990s. As regards total factor productivity growth, several EU countries show higher TFP growth rates than the US. DAVERI (2001) emphasizes that while it is difficult to draw clear conclusions with respect to TFP given the residual character of this variable, the time variations of the TFP growth rates are interesting. The five largest countries in the EU had smaller TFP growth in 1996-99 than in the 1980s and the first part of the 1990s. However, TFP growth has increased in Portugal, Greece, Finland, and Ireland over time. This intra-EU difference is not fully understood, although BASSANINI/SCARPETTA and VISCO (2000) argue that the increase in TFP was relatively high in countries with flexible labor markets and less regulated product markets. However, Spain, the UK, and the Netherlands were indeed countries with considerable labor market deregulation in the 1980s and 1990s. One cannot simply rule out that labor market deregulation is important, as a combination of high ICT investment with accelerated structural changes develops towards a dynamic service society. In the services sector we have, however, well-known and very difficult problems in measuring productivity (ARK/MONNIKHOF/MULDER, 1999; BOSWORTH/TRIPLETT, 2000). With respect to the impact of labor one cannot rule out that rising unemployment rates in the 1980s and early 1990s contributed to growth positively via a positive effort effect of those having a job, but the combined effect of labor quantity and labor effort obviously was negative in many EU countries.

According to McMORROW/ROEGER (2001) there has been some acceleration in the growth contribution of ICT in EU countries where the authors distinguish between two cases, namely EU computer price reductions equal to the US and (ii) the case of only a 50% price reduction, which might reflect various impediments to low prices in Europe, including government regulations in Germany forbidding high rebates on products (this law will be abolished in late 2001). At the bottom line the contribution of ICT to potential output growth in the US has been about twice as high in the EU-15. Among EU countries only Ireland, itself strongly shaped by US multinational investment in ICT, showed a growth contribution of ICT exceeding that in the US in the 1990s. Disregarding Ireland only Sweden, Finland, the UK, the Netherlands, and Belgium came within a range of about 2/3 of US figures in 1995-99. With a direct ICT investment effect of almost 1 percentage point in the late 1990s the US clearly benefited from high nominal investment growth in ICT when prices were falling in relative terms.

Tab. D4: ICT Investment Effects, Contribution to Potential Growth in the 1990's (% Points)

	ICT Price Decline in the EU identical to that in the US		ICT Price Decline in the EU = 50% of that of the US	
	1992-1994	1995-1999	1992-1994	1995-1999
Belgium	0.35	0.60	0.35	0.51
Denmark	0.22	0.38	0.22	0.32
Germany	0.25	0.41	0.25	0.35
Greece	0.12	0.21	0.12	0.18
Spain	0.19	0.39	0.19	0.33
France	0.24	0.42	0.24	0.35
Ireland	0.84	1.91	0.84	1.64
Italy	0.25	0.42	0.25	0.36
Netherlands	0.41	0.67	0.41	0.56
Austria	0.24	0.41	0.24	0.34
Portugal	0.25	0.55	0.25	0.47
Finland	0.31	0.63	0.31	0.53
Sweden	0.30	0.68	0.30	0.57
UK	0.35	0.64	0.35	0.54
EU15	0.27	0.49	0.27	0.41
US	0.40	0.87	0.40	0.87

Source: Mc Morrow, K./ Roeger, W.: Potential Output: Measurement Methods, "New" Economy Influences and Scenarios for 2001-2010 – A Comparison of the EU15 and the US; Economic Papers Nr. 150, April 2001, p. 71.

ICT dynamics are certainly not the full story behind the different growth performance of various OECD countries. However, ICT is one crucial element. Moreover, rapid accumulation of ICT capital might facilitate the exploitation of new knowledge and stimulate innovation as well as diffusion. This is particularly interesting when we take a closer look at the development of R&D-GDP ratios in general and at the development of high technology labor productivity. As regards the latter ROEGER (2001) shows that the US has established a clear lead vis-à-vis the EU since 1993, with a lag of about two years the US world market share of high technology exports has increased strongly while that of Germany and Japan has fallen since 1993. Both Germany and Japan no longer benefit from the exceptional position they had in the Cold War era, namely that they were special among the largest five OECD countries in devoting almost their entire R&D resources to civilian projects.

As the following graph shows, there is a general tendency of R&D-GDP ratios to rise in the late 1990s. Sweden is the leader in the OECD with an increase of slightly more than 1 percentage point in the 1990s, bringing the country to a top figure of roughly 4%. Finland also shows a strong increase and reached 3.2% in 1999. Germany's R&D ratio decreased for more than a decade after 1987, and it has increased only modestly from a bottom figure of about 2% in 1996. The US had a slight reduction of the R&D ratio in the early 1990s but increased in the second half of the decade and was close to 2.3% in 2000. The temporary reduction of the US R&D ratio is, however, mainly due to the falling expenditures on military R&D. This also holds for France and the UK which, however, stabilized R&D ratios in the late 1990s. While Canada and Italy were close together in 1989 with slightly less than 1%, Canada has increased its R&D-GDP ratio, while that of Italy has fallen to a very low level of roughly 0.6 in the mid-1990s. Interestingly, both Japan and Korea reached R&D-GDP ratios slightly above that of Germany in the 1990s.

Fig. D1: R&D GDP Ratios in Selected OECD Countries, 1981-2000

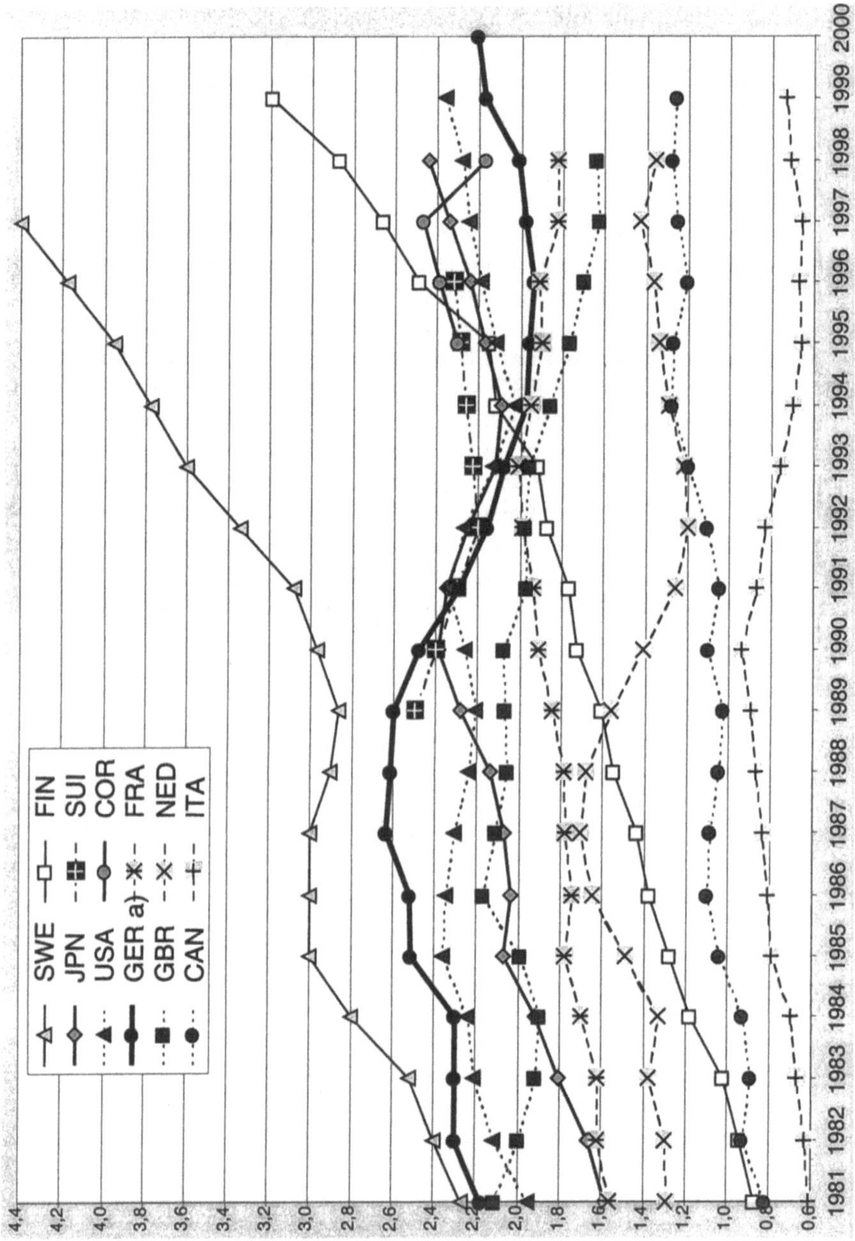

Source: ZEW(2001), based on OECD data.

Given the fact that most EU countries are high wage economies (see the following table), it is obvious that these countries must achieve high labor productivity if full employment is to be reestablished or maintained. High capital intensity (referring here to non-ICT-capital), high ICT intensity, and high knowledge intensity are the three pillars upon which a high wage economy can be built. Facing EU eastern enlargement, one may anticipate that many plants in capital-intensive sectors, e.g. automotive or steel, will be relocated to eastern Europe. Hence it will be all the more important that high wage EU countries generate sufficient increases in R&D capital, human capital, and ICT capital.

Tab. D5: Hourly Wages in Manufacturing Industry (in DM, based on average annual exchange rate)

	1980	2000	% change 1980/2000
Belgium	13,77	23,23	69
Luxembourg	15,63	27,13	74
Sweden	14,47	25,63	77
Netherlands	12,62	22,57	79
France	9,50	18,51	95
Italy	8,03	15,64	95
Kanada	12,73	25,45	100
West Germany	12,93	27,84	115
Switzerland	13,80	30,16	119
Denmark	16,64	36,71	121
Greece	4,36	9,81	125
Norway	14,62	33,13	127
Ireland	8,83	20,33	130
USA	13,20	30,45	131
Finland	9,64	23,30	142
Austria	8,46	20,69	145
United Kingdom	9,89	25,64	159
Japan	6,43	28,49	343
Spain	-	15,06	-
Portugal	-	7,25	-
East Germany	-	19,26	-

Source: Institut der deutschen Wirtschaft, Cologne.

If the expenditures on R&D and software were to systematically increase in OECD countries in the early 21st century this would mean that the share of sunk costs in most products would increase. From a theoretical point of view this reduces price flexibility in the sense that innovative products are typically less exposed to price competition than standardized goods. However, high sunk costs imply that incumbent firms have considerable room to maneuver with respect to temporary price cutting when a newcomer wants to enter the market. They could disregard all sunk costs to fend off the intrusion of newcomers. If newcomers were to try to enter new markets particularly in phases of an economic upswing, this could imply that price increases in the economic upswing will be less pronounced in the future.

2.2 Perspectives on Inflation and Growth

US inflation rates were low throughout the 1990s. This is surprising given the considerable boom in the second half of the 1990s. High economic growth could in itself be an explanation of low inflation to the extent that high growth rates were anticipated in stock markets – we will show a formal model below. Moreover, the strong appreciation of the dollar – in the late 1990s especially vis-à-vis Asian countries – has helped to maintain low inflation rates; this holds at least until 1999/2000 when international oil and gas prices increased strongly. While the previous period of a strong dollar in the mid-1980s was accompanied by a high current account deficit and a high budget deficit, the strong dollar of the late 1990s developed only in combination with a high current account deficit. This current account deficit is not necessarily pointing to competitiveness problems of US industry; rather it seems to be the mirror of high capital inflows in a system of fixed exchange rates: A relatively high marginal product of capital in the US generates high capital inflows; at the same time the low growth rate of Euroland´s three core countries, namely Germany, France and Italy, points to a rather low marginal product of capital in Euroland which is part a broader analysis explaining the weak Euro in 1999/2000 (WELFENS, 2000).

When US stock market indices started to fall in 2000/2001 the euro could not improve its position. Anticipation of changing US long-term interest rates might explain the coincidence of falling stock market prices in the US and a stable dollar; portfolio investors would not switch from the US stock market towards the stock markets in Euroland; rather US investors would move from the US stock market to the US bond market. In both the US and EU-15 a shift in the Phillips curve was observed; that is, unemployment rates could fall without causing inflation rates to rise as steeply as in previous decades; the NAIRU has reduced (RICHARDSON ET AL., 2000). It is unclear whether the shift in the Phillips curve is mainly reflecting changes in goods markets or in labor markets. Part of the explanation might be that the elasticity of the supply side in goods markets has increased in the context of an expanding New Economy; with more and more digital services being developed and sold within the business community and directly to consumers the fact that digital services hardly face any capacity constraint suggests that at least in the services sector the supply elasticity has increased. To the extent that international outsourcing of labor-intensive services is facilitated by the internet there is another argument why tightening labor markets would not lead to wage increases as fast as in the past.

The strong depreciation of the euro in 1999/2000 raises the issue of links between the exchange rate and the price level. A strong depreciation will raise the price of tradables with a certain time lag unless there is a high rate of process innovations which would bring about reductions in unit labor costs (or in capital costs). Even with a depreciation of the currency it remains to be seen how monetary policy will behave on the one hand and how stock markets will develop on the other hand. For the inflation analysis we will focus on a modified quantity equation (WELFENS, 2000):

(I) $MV'(i,Y) = PY + \phi(i,...)P'Q,$

where M is the nominal stock of money, V′ velocity of money, Y real output, P the price level (with $P=(P^T)^{\beta}(P^N)^{1-\beta}$; where ß is the share of tradables in overall demand), ϕ the velocity of the representative portfolio – that is it indicates how often the stock of bonds and equities invested have been sold – P′ the price of stocks, and Q the number of stocks). With the assumption $V= v(i)Y^{\chi}$ – with the absolute value of $0<\chi<1$ – dividing by PY results in the following equation where ϕ is assumed to negatively depend on i and other variables.

(II) $[M/P]\ v(i)Y^{\chi-1} = 1+ \phi(i,...)\ [h],$

The symbol h denotes the value of stocks relative to GDP. Assuming that $\phi[h]$ is relatively small so that we can use the approximation $\ln(1+x)\approx x$ then – assuming ϕ is constant and denoting growth rates by g – we obtain after taking logarithms:

(III) $g_P=g_M - (1-\chi)g_Y - \phi dh/dt.$

If purchasing power parity holds for tradables (T) in the form of $e\ P^{T*} = P^T$ and if we define $P=(P^T)^{\beta}(P^N)^{1-\beta}$, then $g_e + g_{PT*} = g_{PT}$ and therefore $g_P = g_{PN} + \beta(g_e + g_{PT*} - g_{PN})$ so that we obtain the growth rate of tradables prices as the difference between the inflation rate and the change of the relative price ratio $(g_e + g_{PT*} - g_{PN})$ weighted by ß.

According to (III) for a constant nominal interest rate and a given growth rate of the money supply it holds that the inflation rate is the smaller the higher the growth rate of real output and the stronger the increase in the ratio of stock market capitalization relative to output. The coincidence of high growth and exceptionally growing stock market capitalization – relative to output – could indeed explain the inflation puzzle in the US in the 1990s.

Growth Analysis in a Two-Sector Model

Why was growth so high in the US in the 1990s? We consider a two-sector economy so that overall output Y is composed of the ICT sector (sector 1) which accounted for about 10% in the US in 2000 while the EU recorded only about 6%. Denoting the relative price P_1/P_2 as q" we have

(IV) $Y = q''Y_1 + Y_2,$

Assume that sector 2 is the consumption sector then output in terms of consumption units is given by

(IV.I) $g_Y = (Y_1/Y)\,[g_{q''} + g_{Y1}] + (1-(Y_1/Y))\,g_{Y2}$

From this equation it is clear that the relative fall of computer prices reduces overall growth as measured in units of the consumption good. For statistical purposes the real output growth rate is measured in a different way, namely on the basis of constant historical prices:

(IV.I′) $g_Y = (Y_1/Y)\,[g_{Y1}] + (1-(Y_1/Y))\,g_{Y2}$

Here we can directly see that the high growth rate of the US is partly related to the higher share of the ICT sector in overall output and partly due to the high output growth rate of the ICT sector. The US statistical system has made two changes in the 1990s which go beyond the above formula but which almost have no net effect: Introducing a chain-weighted index – reducing the well-known problem of substitution in the presence of relative price changes – for real GDP measurement showed that real GDP growth had to be slightly revised downwards. At the same time the introduction of hedonic pricing which takes into account quality improvement – highly significant in the ICT sector (especially in the case of computers) – brought an upwards revision of real GDP. One can only wonder why Eurostat is not making similar adjustments in its SNA procedures.

We now turn to the analysis of output growth on the basis of factor inputs and total factor productivity (TFP) growth. The US recorded in some years two digit growth rates of total factor productivity growth in the 1990s. Overall TFP is the weighted sum of sectoral TFPs, that in our case is of two sectors; here α_i (i=1,2) is the share of nominal sector 1 output relative to overall nominal output.

(V) $TFP = \alpha_1\,TFP_1 + \alpha_2\,TFP_2$

In a simple growth model based on a production function VF(K,L,Z,Q) with inputs capital K (without computers), labor L and computers Z we can decompose output growth as follows where we use $E_{Y,X}$ to denote partial output elasticities:

(V.I) $g_Y = E_{Y,K}\,g_K + E_{Y,L}\,g_L + E_{Y,Z}\,g_Z + g_V$; here g_V= TFP

(V.II) $g_Y = E_{Y,K}\,g_K + E_{Y,L}\,g_L + E_{Y,Z}\,g_Z + \alpha_1\,TFP_1 + \alpha_2\,TFP_2$

The input factor computer is a proxy for ICT goods used as inputs in firms. Total factor productivity growth in the ICT sector has been enormous in the 1990s in the US and is expected to remain high; while the EU also has had an increase in productivity growth in ICT it is unlikely that West European countries will be on par with the US. It is true that Sweden, Finland, Ireland and some regions in Germany and France have a dynamic ICT industry, but global industry leaders clearly are firms from the US which also spend more than twice as much as the Europeans on software (OECD, 1998). While it is unlikely that the US lead in ICT will continue forever, it might well maintain a considerable lead over several decades.

Next we turn to the issue of heterogeneous labor; that is we will distinguish between unskilled labor L and skilled labor H. The substitution elasticity of L and H with respect to computers is crucial, and here again we might find differences in the US and Europe. However, even if the substitution elasticity in the US would be higher than in the EU high investment in ICT would not necessarily mean a rise of unskilled unemployed in the US. The crucial question is whether the labor market is flexible enough to transitorily allow wage rates of unskilled labor to fall – with employment growth and output growth continuing, finally the demand for unskilled labor will grow strongly; this holds in part because well-paid skilled labor will increasingly demand all kinds of services which are often not very skill intensive.

Next we will slightly modify the production function and assume that infrastructure capital Q enters the production function, too. Assuming a linear-homogeneous function F(...) and a production function Y= QVF(K,L,H,Z), we can write output per unit of skilled labor as follows – with $Y/H := y^h$, $k' := K/H$, $l' := L/H$, $Z/H := z'$ and w, w′ as real wage rate for unskilled and skilled labor and r and p′ denoting the real interest rate and the price of computers, respectively:

$$\text{(VI)} \quad y^h = f(k'(r/w, w'/w, p'/w), l'(r/w, w'/w, p'/w), z'(r/w, w'/w, p'/w), Q, V)$$

$$\begin{aligned}\text{(VII)} \quad d\ln y^h/dt = {} & E_{K/H,r/w}\, g_{r/w} + E_{K/H,w'/w}\, g_{w'/w} + E_{K/H,p'/w}\, g_{p'/w} + \\ & + E_{L/H,r/w}\, g_{r/w} + E_{L/H,w'/w}\, g_{w'/w} + E_{L/H,p'/w} g_{p'/w} + E_{Z/H,r/w}\, g_{r/w} + E_{Z/H,w'/w}\, g_{w'/w} + \\ & + E_{Z/H,p'/w}\, g_{p'/w} + Ey,q' g_Q + Ey',V(\alpha_1\, TFP_1 + \alpha_2\, TFP_2)\end{aligned}$$

Assuming for simplicity that r/w is constant, then the growth rate of output per unit of skilled labor is given by

$$\begin{aligned}\text{(VII')} \quad d\ln y^h/dt = {} & (E_{K/H,w'/w} + E_{L/H,w'/w} + E_{Z/H,\, w'/w})\, g_{w'/w} \\ & + \{[E_{K/H,p'/w} + E_{L/H,p'/w} + E_{Z/H,\, p'/w}] g_{p'/w}\} + Ey',V(\alpha_1\, TFP_1 + \alpha_2\, TFP_2)\end{aligned}$$

Note that the elasticities Ey′,q′, Ey′,v stand for the elasticity of output per unit of skilled labor with respect to infrastructure capital Q and the level of technology V, respectively. In the US the growth rate of $g_{w'/w}$ was positive in the early and mid-1990s; in Euroland it was close to zero. As the three partial elasticities in the first bracket are positive, it is obvious that the skill premium in the US has contributed to growth while for Germany and Euroland, respectively, there was no such growth impact as trade unions managed to maintain or even transitorily reduce vertical wage dispersion. This indeed would have reduced the growth impact of ICT to the extent that expansion of ICT would have required a rising vertical wage dispersion. The direct growth contribution of computers is indicated by the last bracket term{...}. The first two elasticities are positive, and the last one $E_{Z/H,p'/w}$ is negative since the ratio of using computers per unit of skilled labor will fall if computer prices should increase. If the absolute value for this elasticity exceeds $E_{K/H,p'/w} + E_{L/H,p'/w}$ – and this is a realistic case – then the fall of computer prices will go along with a positive value of the term {...}. While government expenditures relative to GDP were 3% p.a. in the US in the 1990s, they were close to 2 % in Germany and only slightly higher in Euroland. If the supply elasticity of infrastructure capital were of equal size in the US and Euroland, the slower growth rate of g_Q in the latter – mainly caused by low infrastructure expenditures in Germany, Belgium and Austria – would explain part of the transatlantic growth differential. Germany together with Belgium indeed recorded the lowest ratio of public investment to GDP in 2000; this might be understandable in Belgium where government is facing the challenge to reduce a debt-GDP ratio of more than 100% in 2000, but for reunited Germany that is a real puzzle since its debt-GDP ratio stood at an uncritical 60%; additionally since economic catching-up of eastern Germany requires relatively high public investment one would expect Germany to have one of the highest public investment-GDP ratios among the members of Euroland.

If we assume for the medium term that the number of skilled labor is exogenous – not a realistic presumption for the long run taking into account long-term opportunities for training and education – then the growth rate Y/H in the above equation is identical with the overall growth rate of output.

2.3 What is the Impact of US Growth Leadership for the Euro?

As we are interested in the long-run link between growth and the exchange rate we will focus on a variant of purchasing power parity:

(1) $eP^* = P$

We assume for nominal demand for money at home and abroad (* for foreign variables) the following two functions where A denotes real wealth and m the real demand for money:

(2a) $M^d = P\, m(Y,i,A),$

(2b) $M^{d*} = P^*\, m^*(Y^*,i^*,A^*),$

Money market equilibrium is given by the nominal money supply M being equal to the nominal money demand.

(3a) $M= M^d$

(3b) $M^* = M^{d*}$

Inserting in equation (1) gives the following equation for the medium term exchange rate e´:

(4) $e' = (M/M^*)\, m^*(Y^*,i^*,A^*)/m(Y,i,A)$

In order to simplify matters we will focus now on a specification which includes only real capital and M/P as assets so that we can state a simple money demand function where K is the real capital stock, P^s is the price of stocks, P the output price, i nominal interest rate and e" the Euler number (WELFENS, 2000):

(2a´) $m = Y\, e''^{-\sigma i}\, KP^s/P;$

(2b´) $m^* = Y^*\, e''^{-\sigma^* i^*}\, K^* P^{s*}/P^*$

If we assume for simplicity that the semi interest elasticity of money demand $\sigma = \sigma^* > 0$ we get:

(4´) $e = (M/M^*)/[(Y/Y^*)(K/K^*)(P^s/P^{s}{}^*)(P/P^*)\, e''^{-\sigma(i-i^*)}]$

Denoting the population as L and output per capita y =Y/L we can restate the equation as follows:

(5) $e'P^*/P = (M/M^*)(y^*/y)(L^*/L)(K^*/K)(P^{s*}/P^s)\, e''^{\sigma(i-i^*)}$

Equation (5) incorporates the Balassa-Samuelson effect in the variable y*/y; this effect suggests that there will be a real appreciation if per capita income y is rising relative to the rest of the world (y*). Taking into account additionally interest rate parity (with a^E as the exogenous expected devaluation rate)

(6) $i = i^* + a^E$

we then can logarithmize equation (5) and insert it into (6) and obtain

(7) $i - i^* = (-1/\sigma)\ln\{[(M/M^*)(y^*/y)(L^*/L)(K^*/K)(P^{s*}/P^s)]/(eP^*/P)\} = a^E$

In principle the devaluation rate can be modeled in various ways which, however, will not be studied here.

Turning back to equation (5) we note the advantage that the left-hand side refers to the real exchange rate so that for the medium term – with P and P* given – the equation can be used to derive e.

Equation (5) says that there will be a real devaluation if per capita money supply grows relative to that abroad or if the real per capita income is falling relative to the value abroad or if the foreign capital stock rises relative to the corresponding figure abroad or if the foreign stock market price is rising relatively to the domestic stock market price. With respect to the euro exchange rate it follows that one should anticipate a nominal and real devaluation of the euro, respectively, if:

- the growth rate of the nominal money supply in Euroland exceeds that of the US
- per capita income growth in the US exceeds that in Euroland
- the growth rate of the US population is higher than in Euroland – which was the case in the 1990s and is likely to remain so for several decades.
- the stock of capital (possibly including R&D capital and public infrastructure capital) grows faster in the US than in Euroland
- the US stock market price exceeds that of Euroland (or falls more slowly than in Euroland)
- nominal interest rate in the US is higher than in Euroland.

Except for the first point all other points indeed could be observed in the US in the 1990s and in particular in 1999/2000 when the euro was launched.

With respect to equation (1) we additionally could study a simple modification which is adequate in a Schumpeterian setting with heterogeneous traded goods and heterogeneous labor – where population L consists of skilled labor H and unskilled labor L′:

$$(1') \quad eP^* = \lambda P$$

The variable λ reflects the heterogeneity of goods; it will exceed unity if the degree of product innovations abroad is higher than at home. If the speed of product innovations should accelerate abroad relative to the home country we will have $d\lambda/dt>0$. The nominal devaluation rate will be

$$(1'') \quad g_e = g_\lambda + (g_P - g_P{}^*)$$

Turning to equation (5) we then have the following modification if one assumes that λ will positively depend on relative cumulated R&D expenditure F^*/F, the ratio of web host densities (containing innovative digital services) Ω^*/Ω and relative human capital stock H^*/H:

$$(5') \quad eP^*/P=\lambda(\Sigma F^*/F; \Omega^*/\Omega, H^*/H)\,[M/M^*][y^*/y][(L^*/L][K^*/K][P^{s*}/P^s]e^{\text{"}\sigma(i-i^*)}$$

From a growth theoretical perspective and from a growth policy perspective one will have to explain why there is a long-term differential in F^*/F, Ω^*/Ω and H^*/H, respectively. From a theoretical perspective there are mainly five points to be considered:

- R&D intensity: The ratio of sectoral R&D relative to value-added in fast growing sectors critically contributes to total factor productivity growth. To the extent that there are sectoral spillover effects one also has to take a closer look at the overall R&D-GDP ratio. The US has a much higher R&D-output ratio in the ICT sector than Germany and Euroland, respectively.

- Trade intensity: Rising export ratios could stimulate innovation dynamics and organizational restructuring in the tradables sector. Moreover, rising import penetration might stimulate the management of the respective sectors to come up with productivity-enhancing restructuring and product and process innovations.
- International differences in human capital intensity, including differences in retraining and training of workers.
- Differences in the expansion of the internet where both the user side (user density, average time spent on the net) and the supply side (host density) are important. The internet and the telecommunications system are crucial for rapid diffusion of knowledge and for the expansion of novel digital services.
- Differences in unemployment: The larger the unemployment rate the lower will be the savings ratio – assuming that savings of unemployed is zero in the long run – and the more widespread could be resistance against the introduction of new technologies.

Comparing the US and Germany/Euroland would require identifying the relevant differences for each point and evaluating the significance for transatlantic growth differentials. There is, however, no doubt that Euroland´s high unemployment rates in Germany, Italy and France are part of Euroland´s rather weak growth performance. Depreciation rates on human capital are certainly considerable especially in all those EU countries which have a high share of long-term unemployed.

3. Some Long Term Aspects

In the US productivity growth in the 1980s and 1990s has benefitted from growing trade – indeed mainly from rising import penetration as is shown in the empirical analysis of MANN (1998); she also shows that Germany had no significant productivity-enhancing effect from trade. This might reflect lack of flexibility of firms in Germany´s tradable sector, but it also could indicate the different regional trade orientation of the US and Germany over time. While the US has strongly increased imports from Asian NICs which increasingly have realized high R&D-GDP ratios and thus can be expected to export more technology-intensive products over time, Germany´s imports from Asia have been rather modest.

A strange finding for Germany also concerns the fact that labor productivity in the high technology sector was lower than average in the 1990s (GRIES/JUNGMITTAG/WELFENS, 2001). This might point both to weaknesses of Germany´s high technology firms and to inefficient R&D subsidization. From this perspective Germany seems to be rather poorly positioned to exploit growing high technology dynamics in the OECD. One may also point out that the R&D expenditure-GDP ratio reached a historical peak of 2.9% in 1989 but then fell – when Theo Waigel was Minister of Finance – to 2.3% in 1998; this ratio has not improved in 1999/2000; and the new Minister of Finance, Hans Eichel, seems to be determined to pursue the benign neglect attitude for R&D promotion. By contrast, the OECD's leading country in terms of R&D-GDP ratio, Sweden, recorded a ratio of 3.8% in 1999. Germany's expenditure on education also is meager: 4.7% in 1998 is slightly more than half the top figures of roughly 8% in Scandinavian countries and also much worse than the US with about 6%. Add to this the fact that Germany ranks – according to OECD analysis focusing on employment, output, exports and R&D in ICT – among the lower third in ICT among the 29 OECD countries.

Germany also is facing problems in the telecommunications sector where the former monopoly operator – under the supervision of the Ministry of Finance – has not been fully privatized. This is much in contrast to the UK, Spain, Italy and several other OECD countries. In addition the internet subsidiary of Deutsche Telekom has dropped its offer of a flat rate for normal telephone users (narrow band and ISDN) as of January 2001. This is much to the disadvantage of a full exploitation of digital growth opportunities; to make matters worse, Deutsche Telekom has offered as a "substitute" DSL which not only is not available at short notice but cannot be used in most parts of eastern Germany where modern glass fiber networks imply the impossibility of benefitting from DSL. The DSL technology is a kind of turbo for fixed networks based on standard twisted copper lines. Thereby Deutsche Telekom is directly undermining the growth potential of eastern Germany which is facing problems in keeping up with west German growth figures.

4. Internet, Telecommunications and Innovation

Following the distinction developed by CASPER/GLIMSTEDT (2001, p. 267), we can distinguish between four layers in the internet:

- (A4) Application layer: eg e-mail, file-transfer protocol, software platforms for B2B and B2C e-commerce, online information
- (A3) Navigation and middleware layer: eg www browsers, search engines and electronic payments systems
- (A2) Connectivity layer: eg internet access, web server parks
- (A1) Network layer: eg trunk networks, fixed local access networks, radio networks, LANs

Both regulation of telecommunications and standardization in the telecommunications market will affect the internet. The network layer is directly linked to access to telecommunications: fixed network, cable TV or power line; mobile access - including satellite access – is partly related to the connectivity layer. The authors try to differentiate between sectoral innovation systems where they partly draw on BRESCHI/MALERBA (1997).

CASPER/GLIMSTEDT (2001) argue that layers (A1) and (A2) are dominated by large integrated network equipment manufacturers (and network operators); in (A2) mainly large network operators are active due to sunk costs of network provision. (A3) and (A4) are characterized more by entrepreneurial software firms. The authors emphasize in their analysis that institutions matter to some extent for innovation in the internet sector and for certain layers, but that technological strategies of dominant firms are often equally important. As regards the role of institutions they point to the importance of flexible labor markets for the success of the high-technology cluster in Silicon valley where many young innovative firms compete with each other; when one of the high-tech start-ups is failing, the clustering of innovative firms makes it relatively easy for highly skilled workers to move to a new firm, expanding on the basis of technology and knowledge.

Fig. D2: Interdependency of Telecommunications and the Internet

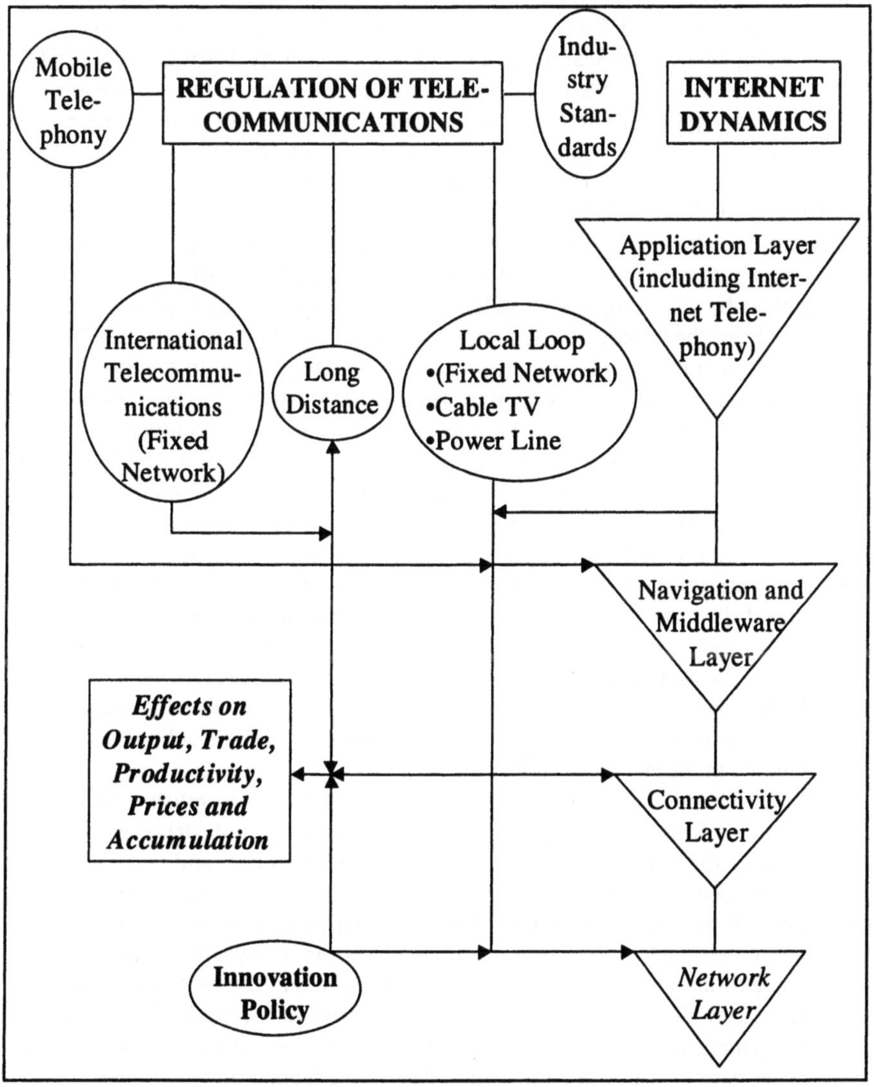

CASPER/GLIMSTEDT (2001, p.277) point to ERICSSON as a well known large company which sets – jointly with other firms – standards in digital wireless systems: "Network-layer firms, by successfully managing the development of new standards, can dramatically lower standard-related risks for entrepreneurial middleware firms...we suggest that technological leadership, when combined with facilitative personnel policies and the willingness to allow external firms to use emerging standards, protocols, and development languages at low costs, can drive the formation of technologically intensive clusters of middleware firms within 'coordinated' types of political economies such as Germany, Japan, or Sweden."

According to the authors, ERICSSON offered long-term stable career perspectives for top graduates in Sweden, resulting in their maintaining a workforce of the best ICT engineers from Scandinavia (or for the Finnish rival Nokia). ERICSSON had to realize in the 1990s that data-communication networking equipment had begun to converge with traditional telecommunication devices. The increased use of IP-based switching forced ERICSSON to adopt standards for data-communication networks. Lacking basic IP skills – especially when compared to the US rival CISCO – ERICSSON was stimulated to forge links with outside developers via alliances and joint ventures. The large network equipment firm thus tried to integrate external sources of innovation: the "string of pearls strategy" as it was dubbed by the company´s strategists. The company´s management learned quickly – from the experience in promoting global standards such as GSM – that use of open data-communication network architectures helped to exploit network externalities; and network externalities are important for establishing a dominant standard. ERICSSON has been an important sponsor and developer of several crucial wireless connectivity standards, concerning for instance LAN, WAP (for transfer of internet content) or Bluetooth for short-distance communication. ERICSSON realized that helping start-ups in the field of wireless technology would contribute to nurture the company´s favored technologies. It even encourages engineers to leave the company and become entrepreneurs, however, they can return if the project fails. As most engineers creating a company remain in the Stockholm area they effectively remain involved in the development of ERICSSON-sponsored standards and most new technologies will be compatible with the company`s standards; the large company in turn is able to integrate the external innovations into its systems, and it benefits via equity holdings and inside links with the managers of the start-up companies.

CASPER/GLIMSTEDT (2001, p.279) summarize the Swedish example of ERICSSON as follows: "In sum, a combination of open, industry-specific standards reduces the risk for engineers who 'go entrepreneurial' and leave Ericsson for start-ups because they simply make it relatively safe to leave from a career perspective. Open standards ensure that skills and knowledge investments made by programmers and engineers are portable. This is a key requirement for the establishment of technical communities that characterize Silicon Valley and other high-technology districts. Within normally conservative Swedish labour markets, this employment insurance is a key catalyst for creating extremely active labour markets necessary to sustain competency-destroying technology strategies."

The key conclusion of the authors is that the Silicon valley model of very flexible markets combined with local university research centers and skilled personnel is only working for some layers of the internet sector, resulting in alternative approaches being used for other layers of the internet sector. The Swedish experience with wireless technology hubs shows that more labor market flexibility can develop within rather rigidly regulated labor markets. If European countries were to simply deregulate all labor markets this would destroy prospects for a networked hub system in which large hub companies rely on rather stable and long term employment relations.

5. ICT Dynamics and Economic Policy

The internet and digital telecommunications will change and affect the economic system in the long run. In the late 1990s there were was excessive speculation about New Economy firms, although this internet hype was partly a speculative bubble, it should not make us ignore the fact that there are some crucial changes brought about by ICT dynamics and likewise some special problems associated with the internet, markets for information and regulation of telecommunications. The high level of ICT dynamics raises several major challenges for policymakers:

- Due to the fact that internet adjustment speeds within markets are generally increasing (where we disregard for a moment the innovation-enhancing effect). However, this does not rule out that the difference between superfast financial markets and slow goods and labor markets will increase, which could create more frequent and accentuated overshooting problems along the lines suggested – in principle – by the Dornbusch model. This then leads to a **challenge for stabilization policy** which might concern above all financial markets instability and only indirectly, goods and labor markets.
- The internet creates larger international markets and thus reinforces international interdependency. Therefore there is a reinforced challenge for an **internationalized policy perspective**; the widely used one-country-model is becoming less and less adequate – this also holds true, of course, due to the increasing tendency for regional integration seen in the late 20th and the beginning of the 21st century. The internationalized policy perspective is concerned with in particular stabilization and growth policies, however also competition policy.
- Ever cheaper and more powerful computer networks will stimulate the creation of more widespread and flexible networks – many of them organized within large global companies; thus the role of large companies in many markets will be reinforced, not least because the internet allows large companies to operate in a rather decentralized fashion; that is, the internet reduces (relative) firm-internal transaction costs in some sectors. Moreover, the notorious imperfections of knowledge markets are likely to reinforce tendencies in knowledge intensive production to organize R&D and

production within the firm. Thus policymakers face a **multinationalization challenge**, MNCs will continue to be increasingly influential over many years. The quest for mobile capital will intensify and as a consequence corporate tax rates can be expected to fall in the long run. The internet will also create problems for VAT tax collection, but increased cooperation in the (enlarged) OECD area might help to effectively deal with this problem.

- In the medium term ICT dynamics, in particular falling computer prices, could reinforce the role of smaller firms in computer-intensive sectors: With falling real prices of computer equipment the market entry costs in many sectors are falling, and as the internet facilitates outsourcing many large firms will want to exploit the new opportunities for cutting costs and getting access to specialized knowledge of new firms. Here the challenge for economic policy is to stimulate **creation of new digital firms – which includes start-up financing policies**. There are two problems here: The benefit in terms of reduced market entry costs is falling real computer prices, thus making it rather difficult for firms to come up with adequate collateral for taking loans; moreover, the fact that in many OECD countries software cannot be activated in the balance sheet implies another problem in financing creation and expansion of young firms – this is all the more relevant as the share of software in "broadly defined investment" is increasing in the long run. The role of equity capital and stock markets could thus increase in the long run since bank loans are rather inadequate for financing the creation and expansion of young digital firms. Continental EU countries (and Japan) which rely heavily on universal banking systems to finance company activities, are disadvantaged vis-à-vis the US and the UK.
- ICT dynamics increase price competition to some extent, meaning that firms facing declining margins face pressure to innovate; in fields with rather innovative and technology-intensive or knowledge-intensive products the price elasticity is lower than in markets with standardized products. This should create a premium for skilled workers and scientists. The wage differential in favor of skilled workers could be reinforced if ICT dynamics represent biased technological progress, namely (unskilled) labor-saving progress. The policy challenge then concerns conflicts over income distribution and **redistribution policy**; this may be mainly a problem for countries and regions in which unequal distribution is a major issue (eg Europe, Latin America, Asia). At the same time one may anticipate a higher yield on education – and retraining – resulting in a long term **challenge for education policy**.
- As ICT dynamics stimulate the expansion of knowledge and R&D-intensive sectors, there is a major **challenge for R&D policy**. This challenge not only means to increase government R&D support, but it also could imply broader challenges for modernizing the innovation system, namely to the extent that ICT dynamics brings distinct challenges – eg in the field of internet-related innovative products and services. Digital Schumpeterian dynamics might indeed present some particular problems for policymakers.

- Since the expansion of digital (tele-)communications network is crucial for the internet society, adequate **telecommunications regulatory policy** will continue to play a crucial role. The fixed network and the new mobile networks are two crucial pillars for the communications system.

A crucial challenge for economic policy concerns the interdependency between regulation of telecommunications and internet dynamics.

As regards innovation dynamics and the expansion of new firms CASPER/GLIMSTEDT (2001) argue that one has to distinguish between several technology regimes. The complexity of typical R&D-projects, the scope of network effects and the technology strategy of leading companies play a role. One regime could be characterized by a parallel innovation race of many firms, all of which employ highly skilled labor and advanced ICT. Another regime is better characterized by the role of the leading company as a technology hub which serves as a node for a network which includes many high technology start-up firms.

Germany has rather rigid labor markets and is facing difficulties in implementing the Silicon Valley innovation model (CASPER, 2001). However, Germany has to find a unique answer to the innovation challenge of the digital age: it will have to modify the national innovation system and to develop new forms of international cooperation on the side of industry and government. It is an open question whether Germany's dynamic sectors in the old economy (e.g. automotive sector, chemical industry) are well prepared to exploit innovation dynamics in complementary ICT fields. At the same time there is a challenge namely to what extent Germany's R&D-policy, traditionally supporting mainly large companies, can find efficient ways to promote start-up firms in new fields of economic activity. There are similar challenges in France. In Italy and Spain a rather strong service orientation might create particular opportunities for new digital services, but lack of skilled personnel and rather limited activities of small high technology firms create an environment different to that of Germany and France.

6. Innovation Policy Challenges

Economic policymakers are facing key challenges in innovation policy in the field of ICT and in growth policy:

- International spillover effects are likely to increase in the internet age. This will make it more difficult to refinance R&D government support programs since a considerable part of innovation and output effects will be realized abroad (not in the country of the original innovator). This will mean that foreign countries will benefit from additional tax revenues which were formerly almost fully accrued to the innovating country; within the OECD this need not to be a dramatic problem if there are symmetrical international spillover effects, however this is doubtful and is an issue for theoretical and empirical research. Internalization of international spillover effects from national R&D projects could be envisaged partly on the basis of international R&D promotion and some form of tax revenue sharing. It is unclear to what extent joint international R&D programs – eg in the EU – are feasible and an efficient answer on the side of policymakers. Proposals for tax revenue sharing have to be developed.

- Diffusion rates will accelerate which stimulates R&D and innovation– at least if the growing internationalization of business allows the recovery of R&D costs despite the shortening of innovation cycles. From this perspective the internationalization of the economy will continue. Here the changing scope of international R&D has to be assessed carefully.
- It seems that the long term investment-GDP ratio is falling while the software expenditure-GDP ratio is increasing. This reinforces the problem of start-up financing in digital innovative firms which will find it difficult to come up with adequate collateral. This reduces the role of bank finance and therefore reinforces the need in continental EU countries to strengthen venture capital financing. Here the topic is what continental EU countries can learn from the US and to what extent recent progress in broadening the venture capital basis in some EU countries is of significance.
- Patenting of innovative services is a notorious problem – copyrights and trademarks are not a full equivalent for patenting; here there are major challenges. Some problems can be solved effectively only at the WTO level; in particular one will have decide which digital products fall in the realm of the comprehensive GATT framework as they are considered to be goods, and which digital products are in effect considered as services and therefore fall under the weaker framework of GATS. The WTO has produced several working papers on the issue but a clear decision has not yet been made. The more innovative digital services are, the more important it seems that they would fall under the strict rules and procedures of GATT.
- Exploitation of network effects is crucial in ICT, providing impetus for government to consider requirements for R&D projects to work on the basis of open standards. Comparing programs in EU countries and in the US (and selected US States, eg California, Massachusetts) will be interesting here.
- Secure payments over the internet will remain a key problem for many years to come meaning joint international R&D projects in this field could be useful. It is rather unclear whether the big players in the services sector which have so far not focused particularly heavily on high technology development will effectively cope with the new challenges of the internet – such an open net poses far more difficult problems than closed company networks or special central bank networks currently used for many international transactions in banking.
- The internet sector itself consists of several layers which apparently require differentiated R&D strategies which in turn call for differentiated R&D support. Depending on the network layer large companies might work as an innovation and technology hub; in such cases innovation policy is facing a difficult challenge as it may want to promote the technological capability of a large company, however ends up instead with inefficient indus-

trial policy. Rent-seeking problems could be crucial in this context. The problem is a different one, if the respective internet layer calls more for promoting technology intensive start-up companies – some of which will grow quickly (gazelles). To the extent that clustering is important for long term innovation success there is, however, the question which institutions will decide about preferable locations and whether the political system has a mechanism (such as the biotechnology race" in Germany) to identify, on well-founded principles, success-promising regions which will get strong support from federal government (or the EU – as in the case of Western Europe).

It is also noteworthy that traditional project support of government in many EU countries has had a strong focus on large companies. This raises the problem how government R&D programs have to be tailored, if small start-up companies are to have a fair chance of participating; eg the issue of programs based on tax rebates versus programs using subsidies - the latter being more favorable for large firms with well-organized lobbies. As regards adequate policy options in the New Economy there is a broad future research agenda.

Appendix I: Data from IMF Economic Outlook 2001 (or website)

Tab. D6: Real Growth of GDP

Real GDP	1981-90 (Tenyear-Average)	1991 - 2000 (Tenyear-Average)	1999*	2000*
Developed Countries	3,1	2,4	2,8	2,7
Important Industrialized C.	2,9	2,2	2,6	2,4
USA	2,9	2,7	4.0	4,0
Japan	4,0	1,3	1,0	1,5
Germany**	2,3	1,9	1,4	2,5
France	2,4	1,8	2,5	3,0
Italy	2,2	1,4	1,2	2,4
United Kingdom***	2,7	2,0	1,1	2,4
Canada	2,8	2,4	3,6	2,6
Other Developed Countries	3,7	3,4	3,5	3,6
Spain	3,0	2,4	3,4	3,5
Netherlands	2,2	2,6	2,6	2,5
Belgium	1,9	1,8	1,4	2,5
Sweden	2,0	1,5	3,2	3,0
Austria	2,2	2,2	2,0	2,5
Denmark	2,0	2,4	1,3	1,5
Finland	3,1	2,0	3,6	3,8
Greece****	1,6	2,2	3,3	3,6
Portugal	2,9	2,5	3,0	3,2
Ireland	3,6	6,5	7,5	7,0
Luxemburg	4,6	4,9	3,5	4,4
Industrialized Countries	2,9	2,2	2,6	2,5
European Union	2,4	2,0	2,0	2,7
Euroland	2,4	1,9	2,1	2,8

* Forecast; USA revised for 1999 and 2000

** until 1991 only West Germany data, *** GDP at Market Prices

Source: INTERNATIONAL MONETARY FUND (1999), World Economic Outlook, p. 170.

Tab. D7: The Tax Burden on Low and Middle Wages (Income tax plus social security contributions in 1999 as % of labour costs)

	(1)	(2)	(3)	(4)
B	34.9	41.3	51.2	52.4
DK	14.6	31.0	40.9	40.9
D	31.1	34.5	47.0	47.0
EL	34.3	36.8	35.2	36.5
E	28.4	30.3	332.6	36.2
F	31.5	38.8	40.4	43.5
IRL	-5.2	19.9	21.5	24.7
I	28.2	37.4	44.2	44.7
L	4.7	11.4	30.0	27.9
NL	21.8	34.2	40.3	41.1
A	19.0	31.8	41.7	43.7
P	22.0	26.0	30.3	32.1
FIN	27.6	40.3	43.3	45.4
S	40.9	44.5	48.8	49.7
UK	14.2	23.8	26.2	25.5
US	12.6	24.5	29.3	29.8
JP	14.7	14.7	18.3	18.4

(1) single individual with two children, earning 67 % of the APW (Average wage of production workers).

(2) married couple with two children and a single earner at the APW.

(3) single individual with no child, earning 67% of the APW.

(4) married couple with two children and two earners, with earnings split between the two partners at 100% and 67% of the APW.

Source: European Commission (2001), European Economy, Supplement A, No 1

Tab. D8: Public Expenditure on Education (% of GDP)

	1995	1996	1997				
	Total	Total	Pre-primary + Not Allocated	Primary	Secondary	Tertiary	Total
B (1)	n.a.	n.a.	0.7	1.2	2.7	1.2	5.7
DK	8.0	8.8	1.2	1.8	3.2	1.8	8.0
D(2)	4.8	4.7	0.6	:	3.0	1.1	4.7
EL(3)	2.9	3.1	:	1.1	1.3	0.8	3.2
E	4.9	4.8	0.3	1.2	2.2	0.9	4.6
F	6.0	6.0	0.7	1.2	3.0	1.1	6.0
IRL	5.2	5.0	0.1	1.6	2.0	1.3	4.9
I	4.7	4.9	0.5	1.1	2.2	0.7	4.5
L	4.4	4.3	0.0	1.9	2.1	0.2	4.1
NL	5.2	5.3	0.4	1.2	1.9	1.4	4.8
A	5.6	6.5	0.6	1.3	2.9	1.7	6.4
P	5.8	5.7	0.6	1.7	2.4	1.0	5.7
FIN	7.3	7.4	0.8	1.6	2.3	2.0	6.7
S	7.8	8.0	0.5	2.1	3.2	2.1	7.9
UK	5.2	5.1	0.4	1.1	2.1	1.1	4.7
EU-15	5.2	5.3	0.5	0.9	2.5	1.1	5.0

Includes public institutions and government-dependent private institutions
(1) The data for B are for 1994
(2) The data for D include primary and secondary combined
(3) The data for EL include pre-primary and primary combined
Sources: European Commission (2001), European Economy, Supplement A, No 1

Tab. D9: Employment Rates by Age Group in 1999

	B	DK	D	EL	E	F	IRL	I	L	NL	A	P	FIN	S	UK	EU-15
50-54	60.3	80.4	73.4	60.1	57.2	74.1	62.1	57.2	64.6	70.2	72.7	71.4	78.9	84.2	75.9	69.2
55-59	36.9	70.9	55.1	47.4	44.8	46.8	50.5	36.6	38.2	49.6	41.0	59.1	54.6	77.8	62.1	50.7
60-64	12.9	34.0	19.6	30.4	24.7	10.1	35.9	17.9	12.9	18.6	11.7	43.6	22.2	47.9	35.6	22.3
65-69	3.8	6.2	5.0	11.5	3.9	2.1	14.3	6.2	n.a	5.2	4.9	24.8	4.4	10.7	11.6	6.5

Source: Survey on work forces - Eurostat from report of HLWSP.

Tab. D10: ICT Production Effects, Contribution to TFP Growth over the 1990's (% Points)

	TFP Growth Increase in the EU's ICT Sector identical to that in the US		Two Scenarios for 1995-1998	
			No TFP Growth Increase in the EU's ICT Sector	TFP Growth Increase in EU ICT Sector = 50% of that in the US
	1990-1995	1995-1998		
Belgium	0.16	0.22	0.14	0.18
Denmark	0.04	0.06	0.04	0.05
Germany	0.13	0.19	0.12	0.16
Greece	0.02	0.04	0.03	0.03
Spain	0.09	0.14	0.09	0.12
France	0.14	0.25	0.15	0.20
Ireland	1.09	2.17	1.41	1.79
Italy	0.13	0.19	0.12	0.15
Netherlands	0.18	0.27	0.18	0.22
Austria	0.10	0.18	0.11	0.14
Portugal	0.11	0.22	0.13	0.17
Finland	0.16	0.38	0.25	0.31
Sweden	0.15	0.27	0.17	0.22
UK	0.17	0.33	0.21	0.27
EU15	0.14	0.24	0.15	0.19
US	0.23	0.50		

Source: Mc Morrow, K./ Roeger, W.: Potential Output: Measurement Methods, "New" Economy Influences and Scenarios for 2001-2010 – A Comparison of the EU15 and the US; Economic Papers Nr. 150, April 2001, p. 69.

Appendix II: A Simple Model of an Economy with Information

We will consider an economy which uses information for consumption purposes and also has an information accumulation activity. Assume that society wants to maximize per capita consumption for a given population $L=L_0$. We will subsequently use a set a parameters which all are assumed to be positive. We assume output is produced according to a linear-homogenous production function of n firms. Output of all identical n firms i (1...n) taken together is

$Y = nF^i(K^i, L^i, Z^i)$ where Z is information subject to a logistical diffusion process. The aggregate production function is Y = F(K,L,Z) where Z is characterized by a diffusion process which is the dynamic constraint for our optimization problem.

$$\text{Max} \int_0^T \ln(C)\, e^{-\rho t}\, dt = U \qquad (1)$$

subject to a diffusion equation for information Z:

(C.1) $dZ/dt = aZ(L-Z)$

Therefore

(C.1´) $dZ/dt/Z = a((L/Z) - 1)$

Output in equilibrium must be equal to aggregate demand which is composed of real expenditures on traditional consumption goods C, on stocking up information bdZ/dt, and on using existing information, jZ. Therefore, an additional constraint is, with q denoting the relative price of information in terms of consumption good:

(C.2) $Z\,F(K, L) = C + q\varepsilon dZ/dt + qjZ$

Inserting (C.1´) and (C.2) results in:

(C.3) $dZ/dt = C + \varepsilon\,(dZ/dt/Z) + jaZ(L-Z) = C + \varepsilon\, a[(L/Z)-1] + jaZ(L-Z)$

The supply of new information is assumed to depend on combining capital and labour, that is every combination of K and L in firms results in some new knowledge. There is additionally an "education and research sector" generating new knowledge. That is, we will assume that accumulation of knowledge is governed by the following simple differential equation:

(C.4) $dZ/dt = \alpha F(K,L) + vZ$

Inserting (C.4) in (C.3) gives:

(C.4´) $dZ/dt = \alpha C + \alpha\varepsilon a((L/\!/)-1) + jaZ(L-Z) + vZ$

Therefore the Hamiltonian is

(C.5) $H = \ln C\, e^{-\rho t} + \pi\{\alpha C + \alpha\varepsilon a((L/\!/)-1) + jaZ(L-Z) + vZ\}$

(C.6) $\partial H/\partial C = 0 \Rightarrow (1/C)\, e^{-\rho t} = \pi\alpha$

The optimal consumption C# therefore is

(C.6.1) $\ln c\# = -\ln\pi - \ln\alpha - \rho t$

This gives a modified Hamiltonian

(C.5) $H\# = (-\ln\pi - \ln\alpha - \rho t)e^{-\rho t} + \pi\{\alpha C + \alpha\varepsilon a((L/Z)-1) + jaZ(L-Z) + vZ\}$

(C.7) $dZ/dt = \partial H/\partial\pi = -(1/\pi)e^{-\rho t} + \{\alpha C + \alpha\varepsilon a((L/Z)-1) + jaZ(L-Z) + vZ\}$

(C.8) $d\pi/dt = -\partial H/\partial Z = \pi(\alpha\varepsilon L/Z^2 - jaL + LjaZ - v)$

and therefore

(C.8´) $d\pi/dt = \pi_o\, e^{(\alpha\varepsilon aL/Z2 - jaL + LjaZ - v)t}$

Optimal consumption can be written implicitly as:

(C.9) $C\# = (1/\pi_o)(1/\alpha)\, e^{[v-\rho + jaL(1 - Z - \alpha\varepsilon/(jZZ))]t}$

Let us assume as a terminal condition $C(\infty) = C^{min}$, which is a politically defined minimum consumption. Then we can calculate π_o to be equal to

(C.10) $\pi_o = 1/(\alpha C^{min})$

Optimal consumption will grow over time if the exponent in (C.9) is positive. We will assume for ease of exposition that v exceeds the time preference rate. Then the exponent is always positive if $1\text{-}Z\text{-}a\varepsilon/(jZZ).\geq 0$. Let us take a look at the benchmark case of zero which requires $1 - a\varepsilon/(jZZ) = Z$.

Assuming that $a\varepsilon/(jZZ)$ is a figure x close to zero we can use the approximation $\ln(1+x)\approx x$, that is $-a\varepsilon//j(ZZ) = \ln Z$ which gives after differentiation with respect to Z:

(C.11) $2a\varepsilon//j(ZZZ) = (1/Z)$

That gives a benchmark figure for Z of

(C.12) $Z = (2a\varepsilon/j)^{(-1/2)}$

An interesting trait of our model is that Z will approach L_o in the long term. As Z will approach L over time as $t \rightarrow \infty$ we can rewrite (C.7) for this case as $\{\alpha C + \alpha\varepsilon a((L/Z)-1) + jaZ(L-Z) + vZ\} = L$.

Appendix III: Innovation, Savings and Growth

The neoclassical growth model and various new growth models are well known. However, even new growth models so far do not offer a straightforward link between Schumpeterian processes and growth – except for the GROSSMAN/HELPMAN-link between product varieties and growth.

It is a fact of life that people tend to consume more if there are more product innovations. We capture this basic insight in a simple way, namely by assuming – for an economy with a given population L - that aggregate consumption C(t) positively depends on real income Y and the number of new adopters who are already using the new product. The number of new adopters is described by a logistical function where evolution of adopters is governed by:

(i) $dZ/dt = \alpha Z[(L/Z) - 1]$

Consumption depends on real output and the number of new adopters in society. Note that $Z(t) = L/[1+(L-1)e^{-dt}]$. Consumption is given by

(ii) $C = cY + bdZ/dt$

The parameter b can be interpreted as a relative price which is assumed to be given; for simplicity alternatively b could depend negatively on (Z/L). Taking into account (i) we have:

(iii) $C = cY + b\alpha Z[(L/Z) - 1]$

In an economy with a neoclassical production function ($Y=K^{\beta} L^{1-\beta}$), zero population growth and capital depreciation δ we have capital accumulation determined by the difference between gross investment I and capital depreciation which is proportionate to the capital stock K.

(iv) $dK/dt = I - \delta K$

Savings S is – reflecting the consumption function (s=1-c; b' = (1-c) b/c) – given by

(v) $S = sY + b'dZ/dt$

(v') $S/L = s\, Y/L + b'\alpha\, [Z/L]\, [L/Z-1]$

Clearly, innovation and diffusion are not only raising consumption but also savings. We next assume that α is a positive function of k: The more (information) capital is available the higher the adoption rate. In such an environment it holds true that capital intensity (K/L) will develop - under the equilibrium condition that savings equals investment - according to

(vi) $dk/dt = sk^{\beta} + \{b'\alpha(k)\,[Z/L]\,[L/Z-1]\} - \delta k$

We are not interested in an explicit solution, but we may note that capital accumulation is accelerated by the diffusion term; the message of the traditional neoclassical growth model that a poor country (read: low initial k) will always – assuming $s=s^*$, $\delta=\delta^*$ and identical production functions in country I and II (*for foreign variable, i.e. country II) – grow faster than the rich country, no longer holds true automatically. In a multi-sector model with many staggered innovation projects there will be many staggered diffusion effects, each governed by logistical diffusion functions $dZ_i/dt = \alpha_i Z_i\,[(L/Z_i) - 1]$; expenditures on each product i are given by $p_i\, dZ_i/dt$, for all products we could calculate a weighted sum reflecting the cyclical innovation component of demand. Such an approach can be used to combine growth analysis and business cycles. An interesting setting would be a case where the population is growing or declining over time ($dL/dt \neq 0$).

The internet will change the function $\alpha(k)$ in a way that diffusion is accelerating, this then should stimulate capital formation temporarily. As soon as all households have adopted the novelty – that is Z=L – we have no effect on dk/dt meaning there is no steady state effect as soon as Z=L. The main effect of a change in the rate of savings is that the new steady state is achieved faster than otherwise.

In the context of a model for an open economy we could model foreign capital intensity and per capita output in a similar way. Cyclical impulses are then generated by the interaction of innovation/diffusion abroad and exports and imports, respectively.

A final remark on network effects. Network effects in our simple approach can be modeled by assuming, in a two-sector model, that the diffusion of goods is positively related to the diffusion of good j. One interesting aspect of the digital economy is the ability of some firms to set standards (eg network operators, PC producers) and then encourage other firms to use this standard as part of complementary services (eg software, innovative services based on the internet protocol). Standards are like a technology hub, and the internet protocol is one major new standard. A particular feature of the digital economy is that open standards which allow interconnectivity allow real exploitation of worldwide network effects. Moreover, the world of the internet has as the basis for communication a so-called layered functional model which allows engineers to handle complex technology as a modular task (BALDWIN/CLARK, 1997). The crucial trait of a modular system is that it is composed of modules which can be developed independently but still function as an integrated system provided that the standard established by the internet has been respected in the design of the module.

Fig. D3: Innovation Policy and ICT Dynamics

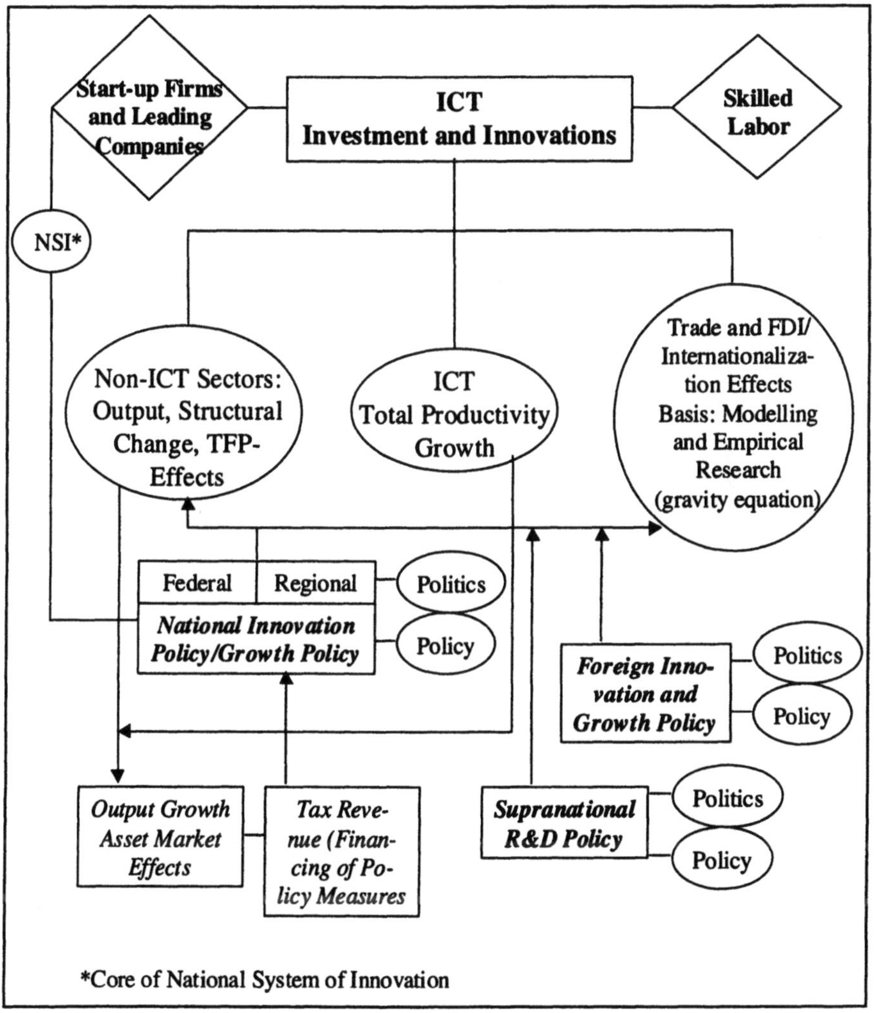

List of Figures

List of Tables

References

ALLEN, D. (1988), New Telecommunication Services, Network Externalities and Critical Mass, Telecommunications Policy, 257-271.

ANALYSYS (1994), Cutting the Cost, Cambridge: Analysys Publications.

ANDERSSON, M. (1996), Full Competition in Telecommunications: The Experience of Finland, paper presented at the international conference Competition in Network Industries: Telecommunications, Energy and Transportation in Europe and Russia, Potsdam, 21-23 November.

AOL (2000), America Online, Inc. 200 Annual Report to Shareholders, Washington DC.

ARK, B. van, MONNIKHOF, E. and MULDER, N. (1999), Productivity in Services: An International Comparative Perspective, Canadian Journal of Economics, Vol. 32, 471-499.

ARNAL, E., OK, W., TORRES, R. (2001), Knowledge, Work Organisation and Economic Growth, Labour Market and Social Policy Occasional Paper No. 50, Paris: OECD.

ARROW, K.J. (1962), The Economic Implications of Learning by Doing, in: The Review of Economic Studies, 1962, 155-173.

AUDRETSCH, D., FELDMAN, M.P. (1996), R&D Spillovers and the Geography of Innovation and Production, American Economic Review, Vol. 86, 630-640.

BALDWIN, C., CLARK, K. (1997), Managing in the Age of Modularity, Harvard Business Review, September-October, 84-93.

BASSANINI, A., SCARPETTA, VISCO, I. (2000), Knowledge, Technology and Economic Growth: Recent Evidence from OECD Countries, mimeo, Paris: OECD.

BAUMOL, W.J., WILLIG, R. (1981), Fixed Costs, Sunk Costs, entry Barriers and Sustainability of Monopoly, Quarterly Journal of Economics, 96, 405-431.

BAYOUMI, T., COE, D.T., HELPMAN, E. (1999), R&D spillovers and global growth, in: Journal of International Economics, 47, 399 – 428.

BECKMANN, M., KÜNZI, H.P. (1982), Mathematik für Ökonomen III, Heidelberg: Springer.

BENHABIB, J., SPIEGEL, M. (1992), The role of human capital in economic development: Evidence from aggregate cross country data, in: Journal of Monetary Economics, 34, 143-73.

BERGMAN, L., DOYLE, C., GUAL, J., HULTKRANTZ, L., NEVEN, D., RÖLLER, L., WAVERMAN, L. (1998), Europe's Network Industries: Conflicting Priorities, London: CEPR.

BERGSTRAND, J. H. (1985), The gravity equation in international trade: Some microeconomic foundations and empirical evidence, in: Review of Economics and Statistics 67, 474-481.

BHF-BANK (2000), „New Economy" spaltet Euro-Kapitalmärkte, BHF-Bank Wirtschaftsdienst, Nr. 1979, June 24, Frankfurt/M.

BLINDER, A.S. (2000), The Internet and the New Economy, Policy Brief, No. 60, Brookings Institution, Washington DC.

BOLTON, P., WHINSTON, M.D. (1991), The Foreclosure Effects of Vertical Mergers, Journal of Institutional and Theoretical Economics, Vol. 147, 207-226.

BOSWORTH, B.P., TRIPLETT, J.E. (2000), What´s New About the New Economy? IT Economic Growth and Productivity, Brookings Institution, mimeo.

BRESCHI, S., MALERBA, F. (1997), Sectoral Innovation Systems: Technological Regimes, Schumpeterian Dynamics, and Spatial Boundaries, in: EDQUIST, C., ed., Systems of Innovation: Technologies, Institutions and Organizations, London: Pinter, 130-155.

BROOKINGS (2000): www.brookings.edu/es/research/ra12summary.htm.

BUDD, A., HOBBIS, S. (1989), Output growth and measure of technology, Discussion paper, London Business School Centre for Economic Forecasting, July 1989.

BUDD, A., HOBBIS, S. (1989a), Cointegration, technology and the long-run production function, Discussion paper, London Business School Centre for Economic Forecasting, July 1989.

CARNOY, M., CASTELLS, M., COHEN, S. S. and CARDOSO, F. H. (1993), The New Global Economy in the Information Age.

CASPER, S. (2001), The German Internet Economy and the Silicon Valley Model: Convergence, Divergence, or Something Else?, in: KOGUT, B. (2002), ed., The Global Internet Economy, forthcoming.

CASPER, S., GLIMSTEDT, H. (2001), Economic Organization, Innovation Systems, and the Internet, Oxford Review of Economic Policy, Vol. 17. 265-281.

CAVES, R.E., FRANKEL, J.A., JONES, R.W. (1990), World trade and payments: an introduction, 5th ed. , Harper Collins.

COE, D. T., MOGHADAM, R. (1993), Capital and trade as engines of growth in France, in: IMF Staff Papers, 40, 542 - 566.

COE, D.T., HELPMAN, E. (1995), International R&D spillovers, in: European Economic Review, 39, 859-887.

COE, S., HELPMAN, E. (1990), Comparative Advantage and Long-Run Growth, in: American Economic Review, 1990, 786-815.

COLECCHIA, A., SCHREYER, P. (2001), ICT Investment and Economic Growth in the 1990s: is the United States a Unique Case? A Comperative Study of nine OECD Countries, STI Working Papers 2001/7, OECD, Paris.

COUNCIL OF ECONOMIC ADVISERS (2000a), Economic Report of the President, Washington DC: Government Printing Office.

COUNCIL OF ECONOMIC ADVISERS (2000b), The Economic Impact of Third-Generation Wireless Technology, White Paper, Washington DC.

COUNCIL OF ECONOMIC ADVISORS (2001), Economic Report of the President, Washington, D.C.

CORDEN, M. (1968), The Theory of Protection, Cambridge: UP.

CREPON, B., HECKEL, TH. (2001), Computerisation in France: An evaluation based on individual company data, Malakof.

DAVERI, F. (2001), Information Technology and Growth in Europe, University of Parma, May 2001, mimeo.

DAVID, P.A., BUNN, J. (1988), The Economics of Gateway Technologies and Network Evolution: Lessons from Electricity Supply History, Information Economics and Policy, 165-202.

DEARDORFF, A. V. (1998), Determinants of bilateral trade: Does gravity work in a neoclassical world, in: FRANKEL, J. A. (Ed.), The regionalization of the world economy, Chicago/London, 7-31.

DEUTSCHE BUNDESBANK (1999), Germany' s Relative Position in the Central and East European Countries in Transition, Monthly Report, October 1999.

DIXIT, A. K., NORMAN, V. (1980), Theory of international trade, Cambridge MA.

DOSI, G., PAVITT, K., SOETE, L. (1990), The Economics of Technical Change and International Trade, New York etc.

DOUGHERTY, C., JORGENSEN, D.W. (1996), Inernational Comparisons of the Sources of Economic Growth, American Economic Review, Vol. 86, 25-29.

DUNNING, J.H. (1981), International Production and the Multinational Enterprise, London: Allen&Unwin.

DURLAUF, S.N., QUAH, D.T. (1998), The New Empirics of Economic Growth, London School of Economics, mimeo, forthcoming in TAYLOR, J. and WOODFORD, M., eds., Handbook of Macroeconomics.

ECONOMIDES, N. (1996), The Economics of Networks, International Journal of Industrial Organization, Vol. 14.

EITO (2000), EITO OBSERVATORY 2000, Frankfurt.

EITO (2000b), EITO Update 2000, Frankfurt/M.

EITO (2001), EITO OBSERVATORY 2001, Frankfurt.

ENGLE, R. F., GRANGER, C. W. J. (1987), Co-integration and error correction: Representation, estimation, and testing, in: Econometrica, 55, 251 - 276.

EUROBIT (1999), Electronic Commerce in the European Union, www.fvit-eurobit.de/pages/eurobit/position/pos038.html.

EUROPÄISCHE KOMMISSION (1999), Leitlinien für die Programme des Zeitraums 2000-2006 *(Guidelines for the programs 2000-2006)*, KOM (1999) 344 endg., 01.07.1999

EUROPEAN COMMISSION (1994), Wachstum, Wettbewerbsfähigkeit, Beschäftigung. Herausforderungen der Gegenwart und Wege ins 21. Jahrhundert. *(Growth, competitiveness, employment. Challenges of the present and roads to the 21st century)*, Luxemburg.

EUROPEAN COMMISSION (1994a), Europe's way to the information society. An action plan, COM (94) 347, Brussels 1994.

EUROPEAN COMMISSION (1995), Green Paper on the Liberalisation of Telecommunications Infrastructure and Cable Television Networks, Brussels.

EUROPEAN COMMISSION (1996), The Institutional Framework for the Regulation of Telecommunications and the Application of EC Competition Rules, final report, Brussels.
EUROPEAN COMMISSION (2000a), European Economy, Supplement A, No. 10/11., Brussels.
EUROPEAN COMMISSION (2000b), The European Economy: 2000 Survey, Brussels.
EUROPEAN COMMISSION (2001), European Economy, Supplement A, No. 7, Brussels.
EUROSTAT (1996), Europa ohne Grenzen, Beilage, Nr. 5.
EVENETT, S. J., HELLER, W. (1998), On theories explaining the success of the gravity equation, Cambridge, M.A., 1998.
FAGERBERG, J. (1988), International Competitiveness, in: Economic Journal, 98, 355 – 374.
FCC (2000), Trends in Telephone Service, March 2000, Washington DC.
FEENSTRA, R.C., MARKUSEN, J.R., HOSE, A.K. (1998), Understanding the home market effect and the gravity equation : the role of differentiating goods, CEPR Discussion Paper no. 2035, London.
FINANCIAL TIMES (2000), Regulator Probe Telecom Loans, London, 29.9.2000, p.1.
FREUND, C., WEINHOLD, D. (2000), On the effect of the Internet on international trade, International Finance Discussion Papers No. 693, Board of Governors of the Federal Reserve System, Washington D.C.
FROOT, K.A., STEIN, J.C. (1991), Exchange Rates and Foreign Direct Investment: An Imperfect Capital Markets Approach, Quarterly Journal of Economics, November, 1191-1217.
FT (2000), Barriers to information 'hamper growth in EU', Financial Times, October 16; see also Commercial Exploitation of Europe's public-sector information. Office for Official Publications of the European Communities, Luxembourg, www.cordis.lu/content.
GEHRKE, B., GRUPP, H. (1994), Hochtechnologie und Innovationspotential, Heidelberg.
GORDON (1999), Has the new economy rendered the productivity slowdown obsolete?, Mimeo.
GORDON (2000), Does the "New Economy" Measure up to the Great Inventions of the Past?, Northwestern University, May 1, 2000, forthcoming in Journal of Economic Perspectives.
GORDON, R.G. (1999), Has the "New Economy" Rendered the Productivity Slowdown Obsolete?, Northwestern University, revised version, June 14, 1999, www.econ.northwestern.edu/faculty-frame.html.
GRAACK, C. (1996), Internationale Aspekte der Telekommunikationswirtschaft: Liberalisierung, Internationale Tarifmechanismen und Wohlfahrtseffekte, EIIW, Potsdam, forthcoming.
GRAACK, C. (1996), Structure of the Telecoms Sector and Degree of Internationalization in Europe and Russia, Discussion Paper No. 17, European Institute for International Economic Relations/University of Potsdam, Potsdam.

GRAACK, C. (1997), Telekommunikationswirtschaft in der Europäischen Union : Innovationsdynamik, Regulierungspolitik und Internationalisierungsprozesse, Heidelberg: Physica.

GRIES, T., JUNGMITTAG, A., WELFENS, P.J.J., eds. (2001), Wachstumsdynamik und Wachstumspolitik in den USA und Europa, Heidelberg: Springer.

GROSSMAN, G., HELPMAN, E..: Innovation an Growth in the Global Economy, MIT Press, Cambridge Mass. 1991.

GRUPP, H., MÜNT, G., SCHMOCH, U. (1994), Trade in High-Technology Products and Patent Statistics - Leading Edge Versus High Level Technology, paper presented at the 1994 Münster Conference of the International Joseph Schumpeter Society: Economic Dynamism: Analysis and Policy, August 17-20, mimeo.

HABUDA, J., STOCK, W.G. (1994), Telekommunikations- und Informationsstruktur in Ungarn, in: Ifo Schnelldienst, 7, 28-38.

HART, O.D., TIROLE, J. (1990), Vertical Integration and Market Foreclosure, Brookings Papers on Economic Activity, 205-276.

HAUSMAN, J.A. (1978), Specification tests in econometrics, in: Econometrica, 46, 1251–1272.

HELPMAN, E., KRUGMAN, P. R. (1990), Market structure and foreign trade, Cambridge, MA/London.

HOEKMAN, B.M., P.C. MAVROIDIS (1995), Linking Competition and Trade Policies in Central and East European Countries, in: WINTERS, L.A., ed., Foundations of an Open Economy, London: CEPR, 111-153.

JAFFE, A., TRAJTENBERG, M., HENDERSON, R.: Geographic Localization of Knowledge Spillovers as Evidenced by Patent Citations, in: The Quarterly Journal of Economics, 1995, 577-598.

JONES, C.I. (1998), Introduction to Economic Growth, Cambridge, MIT Press.

JONES, C.I., WILLIAMS, J.C. (1997), Measuring the Social Return to R&D, Working Paper, Stanford University.

JORGENSEN, D.W., STIROH, K.J. (2000), Raising the Speed Limit: US Economic Growth in the Information Age, Economics Department Working Papers No.261, Paris.

JUNGMITTAG, A. (2001), Internationale Innovationsdynamik, Spezialisierungsstruktur und Außenhandel, in: Gries, T., Jungmittag, A., Welfens, P.J.J. (eds.), Neue Wachstums- und Innovationspolitik in Deutschland und Europa, Heidelberg/Berlin (forthcoming).

JUNGMITTAG, A., WELFENS, P.J.J. (1996), Telekommunikation, Innovation und die langfristige Produktionsfunktion: Theoretische Aspekte und eine Kointegrationsanalyse für die Bundesrepublik Deutschland, EIIW Discussion Paper No.20, Potsdam.

JUNGMITTAG, A., WELFENS, P.J.J. (1998): Telecommunications, Innovation and The Long Term Production Funktion: Theoretical Analysis and a Cointegration Analysis for West Germany 1960-1990. European Institute for International Economic Relations (EIIW) Discussion Paper 52, University of Potsdam, www.euroeiiw.de.

JUNGMITTAG, A., WELFENS, P.J.J. (2002), Telecommunications, innovations and the long-term production function: Theoretical aspects and a cointegration analysis for West Germany 1960 – 1990, in: Audretsch, D., Welfens, P. J.J. (Eds.), The New Economy and Economic Growth in Europe and the US, Berlin/Heidelberg/New York.

KATZ, M.L., SHAPIRO, C. (1994), Systems Competition and Network Effects, Journal of Economic Perspectives, Vol. 8, No.2, 93-115.

KENDE, M. (2000), The Digital Handshake: Connecting Internet Backbones, OPP Working Paper Series No.32, FCC, Washington DC.

KILEY, M.T. (2000), Computers and Growth with Frictions: Aggregate and Disaggregate Evidence, Federal Reserve Board, Mimeo.

KLEIN, M., WELFENS, P.J.J., eds. (1992), Multinationals in the New Europe and Global Trade, New York: Springer.

KNIEPS, G. (1988), Theorie der Regulierung und Entregulierung, in: HORN, M., KNIEPS, G., MÜLLER, J., (eds.), Deregulierungsmaßnahmen in den USA: Schlußfolgerungen für die Bundesrepublik Deutschland, Gutachten des Deutschen Instituts für Wirtschaftsforschung Berlin, Baden-Baden: Nomos, 39-42.

KNIEPS, G. (1997), Phasing out Sector-Specific Regulation in Competitive Telecommunications, Kyklos, Vol.50, Fasc.3, S.325-339.

KRISTIAN, A. (2000), Hochbetrieb in der Tiefsee, LANline 3/99, 22-25.

KRUGMAN, P.R. (1979), A model of innovation, technology transfer, and the world distribution of income, in: Journal of Political Economy 87, 253-266.

LAFFONT, J.-J., TIROLE, J. (2000), Competition in Telecommunications, Cambridge: MIT Press.

LANCASTER, K. (1980), Intra industry trade under perfect monopolistic competition, in: Journal of International Economics 10, 151-176.

LINNEMANN, H. (1966), An econometric study of international trade flows, Amsterdam.

LUCAS, R.E. (1988), On the Mechanics of Economic Development, in: Journal of Monetary Economics, 1988, pp. 3-42.

MAIRESSE, J., HALL, B. (1996), Estimating the Productivity of Research and Development in French and United States Manufacturing Firms: An Exploration of Simultaneity Issues with GMM Methods, in: WAGNER, K., van ARK, B., eds., International Productivity Differences: Measurement and Explanations, Amsterdam.

MANKIW, N.G. (2001), The Inexorable and Mysterious Trade-off Between Inflation and Unemployment, Economic Journal, May, 11, C45-C61.

MANN, C.L. (1998), Globalisation and Productivity in the United States and Germany, in: Black, S. (ed.), Globalisation, Technological Change, and labour Markets, Kluwer, Dordrecht, 1998, 17–44.

MANN, C.L., ECKERT, S.E., KNIGHT, S.C. (2000), Global Electronic Commerce, Washington DC: Institute for International Economics.

McMORROW, ROEGER, W. (2001), The contribution of ICT to growth in Europe and the US: A macroeconomic analysis, Working Paper No 147, European Commission, DG II, Brussels.

MITCHELL, B.M., VOGELSANG, I. (1991), Theory of Telecommunication Pricing, Discussion Paper No. 65, Bad Honnef: WIK.

OECD (1995), Communications Outlook 1995, Paris.

OECD (1998), The Software Sector: A Statistical Profile for Selected OECD Countries, DSTI/ICCP/AH(97)4/REV1, Paris.

OECD (1998a), Conference Conclusions, OECD Ministerial Conference "A Borderless World: Realising the Potential of Global Electronic Commerce", 7-9 October 1998, Ottawa, Canada.

OECD (2000), Local Access Pricing and E-Commerce, Paris 2000.

OECD (2000), Measuring the ICT Sector, Paris, Mimeo.

OFTEL (1995), Pricing of Telecommunications Services from 1997, Consultative Document, London.

OLINER, S.D. (1993), Constant-Quality Price Change, Depreciation, and the Retirement of Mainframe Computers, in: FOSS, M.F., MANSER, M.E. and YOUNG, A.H. (eds.), Price Measurement and Their Uses, Chicago, IL: University of Chicago Press, 19-61.

OLINER, S.D., SICHEL, D.E. (1994), Computers and Output Growth Revisited: How Big is the Puzzle?, Brookings Papers on Economic Activity 2, 273-334.

OLINER, S.D., SICHEL, D.E. (2000), The Resurgence of Growth in the Late 1990s: Is Information Technology the Story?, Journal of Economic Perspectives, Vol. 14, 3-22.

ORDOVER, J.A., SALONER, G., SALOP, S.C. (1990), Equilibrium Vertical Foreclosure, American Economic Review, Vol. 80, 127-142.

PATEL, P., SOETE, L. (1988), L'Évaluation des effets économiques de la technologie, in: STI Review, 4, 133–183.

PELZEL, R.F. (2000), Deregulierte Telekommunikationsmärkte, Heidelberg: Springer.

PERRY, M.K. (1989), Vertical Integration: Determinants and Effects, in: SCHMALENSEE, R. and WILLIG, R.D., ed., Handbook of Industrial Organization, Vol. 1, Amsterdam: North-Holland.

POLS, A. (2002) International Information and Communications Technology Markets and Infrastructures: Some Comparative Statistics, in: WELFENS, P.J.J. and AUDRETSCH, D., eds. (2002), The New Economy and Economic Growth in Europe and the US, Springer-Verlag, Heidelberg and New York.

POSNER, R. (1974), Theories of Economic Regulation, Bell Journal of Economics and Management Science, 5(3), 335-357.

REBELO, S. (1990), Long Run Policy Analysis and Long Run Growth, in: Journal of Political Economy, 1990, 500-521.

REGTP (1999), Telecommunications and Postal Services Markets in 1999, Bonn (www.regtp.de).

RICHARDSON, P. et al. (2000), The Concept, Policy Use and Measurement of Structural Unemployment: Estimating a Time Varying NAIRU across 21 OECD Countries, OECD Economics Department Working Paper No. 250, Paris.

RICHTER, W., SEITZ, H., WIEGARD (1996), Taxes and Firm-Related State Expenditures as Locational Factors, Steuern und unternehmensbezogene

Staatsausgaben als Standortfaktoren, in: SIEBERT, H. (ed.), Steuerpolitik und Standortqualität. Expertisen zum Standort Deutschland, Tübingen 1996, 13-47.

RÖGER, W. (2002), Structural Changes and New Economy in the EU and the US, in AUDRETSCH, D. and WELFENS, P.J.J., eds., The New Economy and Economic Growth in Europe and the US, Heidelberg, New York: Springer.

ROHWEDER, H. C. (1989), Wechselkursvariabilität und internationaler Handel – Eine empirische Analyse der räumlichen Handelsverflechtungen im Rahmen eines TOBIT-Regressionsansatzes, München.

RÖLLER, L. H., WAVERMAN, L. (1996), Telecommunications infrastructure and economic development: A simultaneous approach, WZB discussion paper FS IV 96 - 16.

RÖLLER, L. H., WAVERMAN, L. (1996), The Impact of Telecommunications infrastructure on economic development, in: Howitt, P. (Ed.), The implications of knowledge-based growth for micro-economic policies, Calgary.

ROMER, P.M. (1990), Endogenous Technological Change, in: Journal of Political Economy, 1990, S71-S102.

ROMER, P.M. (1986), Increasing Returns and Long-Run Growth, in: Journal of Political Economy, 1986, 1002-1037.

SAVIOTTI, P.P. (1990), The Role of Variety in Economic and Technological Development, in: SAVIOTTI, P.P. und METCALFE, J.S. (eds.), Evolutionary Theories of Economic and Technological Change: Present Status and Future Prospects, Reading.

SCHMID, B., ZIMMERMANN, H. D. (1997), An Electronic Markets architecture based on a generic concept for electronic product catalogues, University of St. Gallen.

SCHREYER, P. (1998), Information and Communication Technology and the measurement of Real Output, Final Demand and Productivity, STI Working Papers 1998/2.

SCHREYER, P. (2000), The Contribution of Information and Communication Technology to Output Growth: A Study of the G-7 Countries, STI Working Paper 2000/2, Paris.

SCHWARTZ, M. (1997), Telecommunications Reform in the United States: Promises and Pitfalls, in: WELFENS, P.J.J. and YARROW, G. (eds.), Telecommunications and Energy in Systemic Transformation, Heidelberg: Springer.

SHAMPINE, A. (2001), Determinants of the Diffusion of U.S. Digital Telecommunications, Journal of Evolutionary Economics, Vol. 11,249-261.

SICHEL, D.E. (1997), The Computer Revolution: An Economic Perspective, Washington DC: The Brookings Institution.

SICHEL, D.E. (1999), Computers and Aggregate Economic Growth, Business Economics, Vol. XXXVI, No.2, April, 18-24.

SINCLAIR, A. (1996), Vertical Integration in the Electricity Supply Industry: Competition and Investment Issues, paper presented at the international conference Competition in Network Industries: Telecommunications, Energy and Transportation in Europe and Russia, Potsdam, 21-23 November.

SPINDLER, J. (1999), Full competition in telecommunications: The US perspective, in: WELFENS, P. J. J., YARROW, G., GRINBERG, R., GRAACK, C. (eds.), Towards competition in network industries, Berlin/Heidelberg/New York, 123 - 129.

STIGLER, G.J. (1975), The Citizen and the State: Essays on Regulation, Chicago: Chicago University Press.

STIROH (2001), Information Technology and the U.S. Productivity Revival: What Do the Industry Data Say?, Federal Reserve Bank of New York, Mimeo.

STIROH, K.J. (1998a), Computers, Productivity and Input Substitution, Economic Inquiry, Vol. XXXVI, No.2, April, 175-191.

STIROH, K.J. (1998b), Long-Run Growth Projections and the Aggregate Production Finction: A Survey of Models Used by the US Government, Contemporary Economic Policy, Vol:XVI, October, 467-479.

STIROH, K.J. (1999), Is There a New Economy?, Challenge, July-August, 82-101.

TILLY, R., WELFENS, P.J.J., eds. (1996), European Integration as a Challenge to Industry and Government, New York: Springer.

TRIPLETT, J.E., BOSWORTH, B.P. (2000), Productivity in the Services Sector, paper prepared for AER conference, Jan. 7-9, Boston, Mass., Brookings, Washington DC, mimeo.

UPPENBERG, K., VANHOUDT, P. (2000), New Technologies, Measurement Issues and Recent E-conomic Performances, paper presented at the Conference European Economic Growth: The Impact of New Technologies, European Investment Bank, Luxembourg, January, 18, 2001.

VICKERS, J. (1995), Competition and Regulation in Vertically Related Markets, Review of Economic Studies, Vol. 62, 1-17.

VON WEIZSÄCKER, C.C. (1982), Staatliche Regulierung – positive und normative Theorie, in: Schweizerische Zeitschrift für Volkswirtschaft und Statistik, Heft 3, 242-325.

WATERSON, M. (1993), Vertical Integration and Vertical Restraints, Oxford Review of Economic Policy, Vol. 9, 41-57.

WELFENS, P. J. J., GRAACK, C. (1996), Telekommunikationswirtschaft – Deregulierung, Privatisierung und Internationalisierung, Berlin/Heidelberg.

WELFENS, P.J.J. (1992), Market-oriented Systemic Transformations in Eastern Europe, New York: Springer.

WELFENS, P.J.J. (1995), Telecommunications and transition in Central and Eastern Europe, in: Telecommunications Policy, Vol. 19, No. 7, 561-577.

WELFENS, P.J.J. (1996), Telecommunications and Transition in Central and Eastern Europe, in: Telecommunications Policy, Vol. 19, No. 7, 561-577.

WELFENS, P.J.J. (1997), Telecommunications in Systematic Transformation: Theoretical Issues and Policy Options, in: WELFENS, P.J.J., YARROW, G., eds, (1997), Telecommunications and Energy in Systemic Transformation, Heidelberg and New York.

WELFENS, P.J.J. (2000), European Monetary Union and Exchange Rate Dynamics: New Approaches and Application to the Euro, Heidelberg and New York: Springer.

WELFENS, P.J.J. (2001), Aggregation in a Two-Sector Growth Model: A Modified Solow Approach with Cobb-Douglas Production Functions, EIIW-Discussion Paper No. 87, see also www.euroeiiw.de

WELFENS, P.J.J. (2001), Stabilizing and Integrating the Balkans. Economic Analysis of the Stability Pact, EU Reforms and International Organizations, Berlin/Heidelberg/New York.

WELFENS, P.J.J. (2001), Wachstum, Produktivität und Innovationsdynamik in den USA, Europa und Japan, in: GRIES, T., JUNGMITTAG, A., and WELFENS, P.J.J., eds., Wachstumsdynamik und neue Wachstumspolitik in Europa und USA, Heidelberg: Springer (forthcoming).

WELFENS, P.J.J. (2001), Growth and Instability in Open Digital Economies, download under www.euroeiiw.de (forthcoming)

WELFENS, P.J.J., YARROW, G., eds. (1996), Telecommunications and Energy in Transforming Economies. International Dynamics, Deregulation and Adjustment in Network Industries, Heidelberg: Springer.

WELFENS, P.J.J., GRAACK, C. (1996a), Telekommunikationswirtschaft. Deregulierung, Privatisierung und Internationalisierung, Heidelberg: Springer.

WELFENS, P.J.J., GRAACK, C. (1996b), Telekommunikationsmärkte in Europa: Marktzutrittshemmnisse und Privatisierungsprobleme aus Sicht der Neuen Politischen Ökonomie, Discussion Paper No. 22, European Institute for International Economic Relations, Potsdam.

WELFENS, P.J.J., YARROW, G., eds, (1997), Telecommunications and Energy in Systemic Transformation, Heidelberg and New York.

WELFENS, P.J.J., AUDRETSCH, D., ADDISON, J., GRUPP, H. (1998), Technological Competition, Employment and Innovation Policies in OECD Countries, Heidelberg and New York: Springer.

WELFENS, P.J.J., AUDRETSCH, D., ADDISON, J., GRIES, T., GRUPP, H. (1999), Globalization, Economic Growth and Innovation Dynamics, Heidelberg and New York: Springer.

WELFENS, P.J.J., GRAACK, C. (1996), Telekommunikationswirtschaft, Deregulierung, Privatisierung und Internationalisierung, Heidelberg: Springer.

WELFENS, P.J.J., JUNGMITTAG, A. (2000), Auswirkungen einer Internet Flat Rate auf Wachstum und Beschäftigung in Deutschland, EIIW-Discussion Paper No. 52, see also www.euroeiiw.de

WELFENS, P.J.J., JUNGMITTAG, A. (2000), Effects of an Internet Flat Rate on Growth and Employment in Germany, European Institute for International Economic Relations, www.euroeiiw.de

WELFENS, P.J.J., JUNGMITTAG, A. (2001), Internetdynamik, Telekomliberalisierung und Wissensgesellschaft, Heidelberg and New York: Springer.

WELFENS, P.J.J., YARROW, G., GRINBERG, R., GRAACK, C., YARROW, G., eds. (1999), Towards Competition in Network Industries, Heidelberg and New York.

WELLENIUS, B. et al. (1993), Telecommunications. World Bank Experience and Strategy, World Bank Discussion Paper, No. 192, Washington DC.

WHELAN, K. (2000a), Computers, Obsolescence, and Productivity, Federal Reserve Board, Finance and Economics Discussion Series, No. 6.

WHELAN, K. (2000b), A Guide to the Use of Chain-Aggregated NIPA Data, Federal Reserve Board, Finance and Economics Discussion Series, No. 35.

WIELAND, B. (1995), Telekommunikation und vertikale Integration am Beispiel des Bankensektors, Heidelberg.

WILLIAMSON, O. (1985), The Economic Institutions of Capitalism, Free Press, New York.

YARROW, G. (1991), Vertical Supply Arrangements: Issues and Applications in the Energy Industries, Oxford Review of Economic Policy, Vol. 7, 35-53.

YARROW, G. (1994), Privatization, Restructuring, and Regulatory Reform in Electricity Supply, in BISHOP, M., KAY, J., MAYER, C., eds., Privatization and Economic Performance, Oxford: Oxford University Press.

YPSILANTI, D., KELLY, T. (1993), Fostering Telecommunications Development. The Role of the OECD, in: STEINFIELD, C., BAUER, J.M., L. CABY, eds., Telecommunications in Transition, London: Sage, 118-139.

Further Publications by *Paul J. J. Welfens*

P. J. J. Welfens
Internationalisierung von Wirtschaft und Wirtschaftspolitik
Eine Analyse der Dynamik und Gestaltbarkeit von Wirtschaft und Politik in einer sich wandelnden Weltwirtschaft
1990. XV, 443 Pages. 57 Figs., 47 Tab.,
Softcover, ISBN 3-540-52511-4

P. J. J. Welfens
Market-oriented Systemic Transformations in Eastern Europe
Problems, Theoretical Issues, and Policy Options
1992. XII, 261 Pages. 20 Figs., 29 Tab.,
Hardcover, ISBN 3-540-55793-8

M. W. Klein, P. J. J. Welfens (Eds.)
Multinationals in the New Europe and Global Trade
1992. XV, 281 Pages. 24 Figs., 75 Tab.,
Hardcover, ISBN 3-540-54634-0

P. J. J. Welfens
Grundlagen der Wirtschaftspolitik
1995. XVIII, 529 Pages. 57 Figs., 7 Tab.,
Softcover, ISBN 3-540-59289-X

R. Tilly, P. J. J. Welfens (Eds.)
European Economic Integration as a Challenge to Industry and Government
Contemporary and Historical Perspectives on International Economic Dynamics
1996. X, 558 Pages. 43 Figs.,
Hardcover, ISBN 3-540-60431-6

P. J. J. Welfens (Ed.)
Economic Aspects of German Unification
Expectations, Transition Dynamics and International Perspectives
2nd revised and enlarged edition
1996. XV, 527 Pages. 34 Figs., 110 Tab.,
Hardcover, ISBN 3-540-60261-5

P. J. J. Welfens
European Monetary Integration
EMS Developments and International Post-Maastricht Perspectives
3rd revised and enlarged edition
1996. XVIII, 384 Pages. 14 Figs., 26 Tab.,
Hardcover, ISBN 3-540-60260-7

P. J. J. Welfens, C. Graack
Telekommunikationswirtschaft
Deregulierung, Privatisierung und Internationalisierung
1996. XVIII, 272 Pages. 41 Figs., 18 Tab.,
Hardcover, ISBN 3-540-60690-4

P. J. J. Welfens (Ed.)
European Monetary Union
Transition, International Impact and Policy Options
1997. X, 467 Pages. 50 Figs., 31 Tab.,
Hardcover, ISBN 3-540-63309-7

P. J. J. Welfens, G. Yarrow (Eds.)
Telecommunications and Energy in Systemic Transformation
International Dynamics, Deregulation and Adjustment in Network Industries
1997. XII, 501 Pages. 39 Figs.,
Hardcover, ISBN 3-540-61586-5

P. J. J. Welfens, H. C. Wolf (Eds.)
Banking, International Capital Flows and Growth in Europe
Financial Markets, Savings and Monetary Integration in a World with Uncertain Convergence
1997. XIV, 458 Pages. 22 Figs., 63 Tab.,
Hardcover, ISBN 3-540-63192-5

P. J. J. Welfens, D. Audretsch, J. T. Addison, H. Grupp
Technological Competition, Employment and Innovation Policies in OECD Countries
1998. VI, 231 Pages. 16 Figs., 20 Tab.,
Hardcover, ISBN 3-540-63439-8

J. T. Addison, P. J. J. Welfens
Labor Markets and Social Security
Wage Costs, Social Security Financing and Labor Market Reforms in Europe
1998. IX, 404 Pages. 39 Figs., 40 Tab.,
Hardcover, ISBN 3-540-63784-2

P. J. J. Welfens, K. Gloede, H. G. Strohe, D. Wagner (Eds.)
Wirtschaftswissenschaftliche Beiträge, Bd. 169.
Systemtransformation in Deutschland und Russland
Erfahrungen, ökonomische Perspektiven und politische Optionen
1998. XVI, 573 Pages, 69 Figs., 46 Tab.,
Softcover, ISBN 3-7908-1157-2

P. J. J. Welfens
EU Eastern Enlargement and the Russian Transformation Crisis
1999. X, 151 Pages. 12 Figs., 25 Tab.,
Hardcover, ISBN 3-540-65862-9

P. J. J. Welfens
Globalization of the Economy, Unemployment and Innovation
1999. VI, 255 Pages. 11 Figs., 31 Tab.,
Hardcover, ISBN 3-540-65250-7

P. J. J. Welfens, G. Yarrow, R. Grinberg, C. Graack (Eds.)
Towards Competition in Network Industries
Telecommunications, Energy and Transportation in Europe and Russia
1999. XXII, 570 Pages. 63 Figs., 63 Tab.,
Hardcover, ISBN 3-540-65859-9

P. J. J. Welfens, J. T. Addison, D. B. Audretsch, T. Gries, H. Grupp
Globalization, Economic Growth and Innovation Dynamics
1999. X, 160 Pages. 15 Figs., 15 Tab.,
Hardcover, ISBN 3-540-65858-0

P. J. J. Welfens, C. Graack (Eds.)
Wirtschaftswissenschaftliche Beiträge, Bd. 172
Technologieorientierte Unternehmensgründungen und Mittelstandspolitik in Europa
Probleme – Risikokapitalfinanzierung – Internationale Erfahrungen
1999. XVI, 199 Pages. 17 Figs., 22 Tab.,
Softcover, ISBN 3-7908-1211-0

R. Tilly, P. J. J. Welfens (Eds.)
Economic Globalization, International Organizations and Crisis Management
Contemporary and Historical Perspectives on Growth, Impact and Evolution of Major Organizations in an Interdependent World
2000. XII, 408 Pages. 11 Figs., 20 Tab.,
Hardcover, ISBN 3-540-65863-7

P. J. J. Welfens, E. Gavrilenkov (Eds.)
Restructuring, Stabilizing and Modernizing the New Russia
Economic and Institutional Issues
2000. XIV, 516 Pages. 82 Figs., 70 Tab.,
Hardcover, ISBN 3-540-67429-2

P. J. J. Welfens
European Monetary Union and Exchange Rate Dynamics
New Approaches and Applications to the Euro
2001. X, 159 Pages. 26 Figs., 12 Tab.,
Hardcover, ISBN 3-540-67914-6

P. J. J. Welfens
Stabilizing and Integrating the Balkans
Economic Analysis of the Stability Pact, EU Reforms and International Organizations
2001. XII, 171 Pages. 6 Figs., 18 Tab.,
Hardcover, ISBN 3-540-41775-3

P. J. J. Welfens, B. Meyer, W. Pfaffenberger, P. Jasinski, A. Jungmittag
Energy Policies in the European Union Germany's Ecological Tax Reform
2001. VII, 143 Pages. 21 Figs., 41 Tab.,
Hardcover, ISBN 3-540-41652-8

P. J. J. Welfens (Ed.)
Internationalization of the Economy and Environmental Policy Options
2001. XIV, 442 Pages. 57 Figs., 61 Tab.,
Hardcover, ISBN 3-540-42174-2

H.-J. Bunte, P. J. J. Welfens
Wettbewerbsdynamik und Marktbegrenzung auf Telekommunikationsmärkten
Juristisch-ökonomische Analyse und rationale Regulierungsoptionen für Deutschland
2002. XII, 201 Pages. 11 Figs., 2 Tab.,
Softcover, ISBN 3-540-43018-0

P. J. J. Welfens, A. Jungmittag
Internet, Telekomliberalisierung und Wirtschaftswachstum
10 Gebote für ein digitales Wirtschaftswunder
2002. XII, 188 Pages. 129 Figs., 15 Tab.,
Hardcover, ISBN 3-540-42759-7

D. B. Audretsch, P. J. J. Welfens (Eds.)
The New Economy and Economic Growth in Europe and the USA
2002, XII, 350 Pages. 28 Figs., 59 Tab.,
Hardcover, ISBN 3-540-43179-9

MIX
Papier aus verantwortungsvollen Quellen
Paper from responsible sources
FSC® C105338

If you have any concerns about our products,
you can contact us on
ProductSafety@springernature.com

In case Publisher is established outside the EU,
the EU authorized representative is:
Springer Nature Customer Service Center GmbH
Europaplatz 3, 69115 Heidelberg, Germany

Printed by Libri Plureos GmbH
in Hamburg, Germany